The Trachtenberg Speed System of Basic Mathematics

THE Trachtenberg SPEED SYSTEM OF Basic Mathematics

TRANSLATED AND ADAPTED
BY ANN CUTLER
AND RUDOLPH McSHANE

DOUBLEDAY & COMPANY, INC.
GARDEN CITY, NEW YORK

Library of Congress Catalog Card Number 60–13513

Printed in the United States of America
Designed by Alma Reese Cardi

Contents

Foreword

The teacher called on a nine-year-old boy who marched firmly to the blackboard upon which was a list of numbers a yard long. Standing on tiptoe to reach the top, he arrived at the total with what seemed the speed of light.

A small girl with beribboned braids was asked to find the solution of 735352314 times 11. She came up with the correct answer—8088875454—in less time than you can say multiplication table. A thin, studious-looking boy wearing silver-rimmed spectacles was told to multiply 5132437201 times 452736502785. He blitzed through the problem, computing the answer—2323641669144374104785—in seventy seconds.

The class was one where the Trachtenberg system of mathematics is taught. What made the exhibition of mathematical wizardry more amazing was that these were children who had repeatedly failed in arithmetic until, in desperation, their parents sent them to learn this method.

The late Jakow Trachtenberg, founder of the Mathematical Institute in Zurich, Switzerland, and originator of the startling new system of arithmetic, was of the firm opinion that everyone comes into the world with "phenomenal calculation possibilities."

The Trachtenberg method is not only speedy but simple. Once one has mastered the rules, lightning calculation is as easy as reading a story. It looks like magic, but the rules are based on sound logic.

Trachtenberg, a brilliant engineer with an ingenious mind, originated his system of simplified mathematics while spending years in Hitler's concentration camps as a political prisoner. Conceived in tragedy and amidst brutal hardships, this striking work cannot be separated from the life of its originator for it is quite possible that had Professor Trachtenberg's life run a more tranquil course he might never have conceived the system which has eliminated the drudgery so often associated with arithmetic.

The life of Trachtenberg is as fascinating and astounding as his brilliant mathematical system which many experts believe will eventually revolutionize the teaching of arithmetic in schools throughout the world.

A Russian, born in Odessa, June 17, 1888, Jakow Trachtenberg early showed his genius. Graduating with highest honors from the famous Berginstitut (Mining Engineering Institute) of St. Petersburg, he entered the world-renowned Obuschoff shipyards as a student-engineer. While still in his early twenties, he was named Chief Engineer. In those Czar-ruled days, there were ambitious plans to create a superlative navy and 11,000 men were under Trachtenberg's supervision.

Though he headed the Obuschoff shipyards, Trachtenberg was a dedicated pacifist. At the outbreak of World War I he organized the Society of Good Samaritans which trained Russian students to care for the wounded—a work which received special recognition from the Czar.

The murder of the imperial family in 1918 put an end to the Russian dream of a grandiose navy. It also ended Trachtenberg's personal hope of a happy, peaceful life. Hating brutality and violence, Trachtenberg became their victim.

As the Communist horde swept Russia, looting, raping, and killing, Trachtenberg spoke out against the savagery and lawlessness. The criticism imperiled his life. Early in 1919, he learned that he was slated to be murdered. Dressed as a peasant, walking at night, hiding out through the day, he made his way into Germany.

Berlin, with its beautiful wide streets, its cold, sparkling, weather, reminded him of St. Petersburg and became his home. In a tiny room at an unpretentious address, he started life anew and made friends with the bitter, disillusioned young intellectuals of the postwar era. He became their leader. As the editor of a magazine, he often spoke for this group when he urged Germany towards a future of peace.

Trachtenberg married a beautiful woman of the aristocracy. His reputation grew as he wrote a number of critical works on Russia and compiled the first reference book on Russian industry. He was looked upon as Europe's foremost expert on Russian affairs. His inventive mind set itself another task. He devised a method of teaching foreign languages which is still used in many German schools.

The upheaval of his early years seemed to have been left behind. But with the coming of Hitler, Trachtenberg's life once more took on the familiar pattern of strife. Courageously, he spoke out against fascism. Trachtenberg's reputation was such that Hitler at first chose to overlook his attacks. But when Trachtenberg's accusations grew more pointed, Hitler marked him for oblivion.

In 1934, knowing if he remained in Germany he would be liquidated, Trachtenberg once more fled for his life. Accompanied by his wife, he escaped to Vienna where he became editor of an international scientific periodical.

While the world was preparing for war, Trachtenberg, to further the cause of peace, wrote *Das Friedensministerium*

(The Ministry of Peace), a widely read work which brought him the plaudits of such statesman as Roosevelt, Masaryk, and Van Zeeland.

But all over the world peace was dying. The Germans marched on Austria. Trachtenberg's name headed Hitler's most-wanted list. He was seized and thrown into prison.

He managed to escape to Yugoslavia where he and his wife, Countess Alice, lived like hunted animals, rarely venturing out during the day, making no friends or acquaintances. But his freedom was brief. He was awakened one night by the heavy pounding of fists on the door—the Gestapo was calling. Hitler's men had caught up with him.

He was shipped in a cattle car to a concentration camp—one noted for its brutality. The slightest variance from the rules resulted in outrageous forms of punishment. Daily the ranks of the prison were decimated by the ruthlessly random selection of victims for the ovens.

To keep his sanity, Trachtenberg moved into a world of his own—a world of logic and order. While his body daily grew more emaciated, and all about him was pestilence, death, and destruction, his mind refused to accept defeat and followed paths of numbers which, at his bidding, performed miraculous feats.

He did not have books, paper, pen, or pencil. But his mind was equal to the challenge. Mathematics, he believed, was the key to precise thinking. In happier times, he had found it an excellent recreational outlet. In a world gone mad, the calm logic of numbers were like old friends. His mind, arranging and re-arranging, found new ways of manipulating them.

He visualized gigantic numbers to be added and he set himself the task of totaling them. And since no one can remember thousands of numbers, he invented a fool-proof method that would make it possible for even a child to add *thousands* of numbers together without making a mistake—without, in fact, ever adding higher than eleven.

During his long years in the living hell of the concentration camp, every spare moment was spent on his simplified system of mathematics, devising shortcuts for everything from multiplication to algebra. The corruption and misery, the cries from clammy cells and torture chambers, the stench of ovens, the atrocities, and the constant threat of death, faded as he doggedly computed mathematical combinations —reckoning rules, proving and proving again, then starting over again to make the system even simpler.

The hardships acted as a spur to his genius. Lacking paper, he scribbled his theories on bits of wrapping paper, old envelopes, the backs of carefully saved German work sheets. Because even these bits of paper were at a premium he worked everything in his head, putting down only the finished theories.

Today those using the Trachtenberg method find it so easy that all problems can be worked in the head and only the answers put down.

Shortly after Easter in 1944, Trachtenberg learned he was to be executed—the decree had come from above and was no longer speculation or foreboding. Trachtenberg faced the fact, then lost himself in his own world. Calmly he went on working—juggling equations, reckoning formulae, working out rules. He had to get his system finished! To a fellow prisoner, he entrusted his work. He had been in prison almost seven years.

Madame Trachtenberg, who had never been far from the concentration camp, learned of the death sentence. Parting with the last of her jewels and money, she bribed and coerced and managed to have him transferred surreptitiously to another camp just before the sentence was to be carried out.

He was sent to Leipzig which had been heavily bombed

and everything was in a state of chaos. There was no food, no heat, no facilities. In the dismal barracks, the rising tiers of hard bunks were so crowded it was impossible to lie down. Morale had never been so low. Often the dead lay for days, the inmates too weak to dig graves and the guards too panicky to enforce orders.

In the confusion, a determined man, willing to risk his life, could escape to freedom. Trachtenberg took the chance and crawled through the double wire fences in the dead of night. He joined his wife, who had devoted all her time, strength, and money in trying to help him. But Trachtenberg had no passport, nor papers of any kind. He was a stateless citizen, subject to arrest.

Once again, he was taken into custody. A high official who knew of Trachtenberg's work, sent him to a labor camp in Trieste. Here he was put to work breaking rock, but the weather was milder and the guards not so harsh.

Quietly, Madame Trachtenberg bribed guards to take messages to her husband and an escape was again arranged. On a starless night early in 1945, Trachtenberg climbed a wire fence and crawled through the long grass as guards stationed in watch towers shot at him. It was his last escape. Madame Trachtenberg waited for him at the appointed place. Together they made their way across the border to Switzerland.

In a Swiss camp for refugees he gathered his strength. His hair had turned white and his body was feeble, but the years of uncertainty and despair had left him undefeated. His eyes, a clear, calm blue was still valiant. His eagerness and warmth, his intense will to live, were still part of him.

As he slowly convalesced, he perfected his mathematical system which had kept him from losing his mind, which had enabled him to endure the inquisition of the Gestapo, and which now enabled him to start a new life.

It was to children, whom Trachtenberg loved, that he first taught his new and simplified way of doing arithmetic. He had always believed that everyone was born rich in talents. Now he set out to prove it. Deliberately he chose children who were doing poorly in their school work.

These were children used to failure, shy and withdrawn; or the other extreme, boastful and unmanageable. All of them were unhappy, badly adjusted youngsters.

The children's response to the new, easy way of doing arithmetic was immediate. They found it delightfully like a game. The feeling of accomplishment soon made them lose their unhappy traits.

Equally important were the by-products the pupils garnered while learning the new system. As these youngsters became proficient in handling numbers, they achieved a poise and assurance that transformed their personalities and they began to spurt ahead in all their studies. The feeling of accomplishment leads to greater effort and success.

To prove the point that anyone can learn to do problems quickly and easily, Trachtenberg successfully taught the system to a ten-year-old—presumably retarded—child. Not only did the child learn to compute, but his IQ rating was raised. Since all problems are worked in the head, he acquired excellent memory habits and his ability to concentrate was increased.

In 1950, Trachtenberg founded the Mathematical Institute in Zurich, the only school of its kind. In the low, spreading building that houses the school, classes are held daily. Children ranging in age from seven to eighteen make up the daytime enrollment. But the evening classes are attended by hundreds of enthusiastic men and women who have experienced the drudgery of learning arithmetic in the traditional manner. With a lifetime of boners back of them, they delight in the simplicity of the new method. Proudly, they display their new-found mathematical brilliance. It is probably the only school in the world where students—both day and evening—arrive a good half hour before class is called to order.

What *is* the Trachtenberg system? What can it do for *you?*

The Trachtenberg system is based on procedures radically different from the conventional methods with which we are familiar. There are no multiplication tables, no division. To learn the system you need only be able to count. The method is based on a series of keys which must be memorized. Once you have learned them, arithmetic becomes delightfully easy because you will be able to "read" your numbers.

The important benefits of the system are greater ease, greater speed, and greater accuracy. Educators have found that the Trachtenberg system shortens time for mathematical computations by eighty per cent.

All operations involving calculations are susceptible to error whether by human or mechanical operation. Yet it has been found that the Trachtenberg system, which has a unique theory of checking by nines and elevens, gives an assurance of ninety-nine per cent accuracy—a phenomenal record.

The great practical value of this new system is that, unlike special devices and tricks invented in the past for special situations, it is a *complete system.* Much easier than conventional arithmetic, the Trachtenberg system makes it possible for people with no aptitude for mathematics to achieve the spectacular results that we expect of a mathematical genius. Known as the "shorthand of mathematics," it is applicable to the most intricate problems.

But perhaps the greatest boon of this new and revolutionary system is that it awakens new interest in mathematics, gives confidence to the student, and offers a challenge that spurs him on to mastering the subject that is today rated as "most hated" in our schools.

Prof. Trachtenberg believed the reason most of us have difficulties juggling figures is not that arithmetic is hard to comprehend, but because of the outmoded system by which we are taught—an opinion which is born out by many educators.

A year-long survey conducted by the Educational Testing Service of Princeton University revealed that arithmetic is one of the poorest-taught subjects in our schools and noted that there has been little or no progress in teaching arithmetic in this country in the past century; that the important developments that have taken place in mathematical science since the seventeenth century have not filtered down into our grade and high schools. And the results, says the report, are devastating. In one engineering school, seventy-two per cent of the students were found so inadequate mathematically that they had to take a review of high-school mathematics before they could qualify for the regular freshman course.

This is particularly tragic today when there is an urgent need for trained scientists and technicians with a firm grasp of mathematics. The revulsion to mathematics which educators say plays such a strong role in determining the careers of young people, begins at the level of the elementary and secondary schools. It is at this stage that the would-be engineers and scientists of tomorrow run afoul of the "most hated subject." From then on, arithmetic is left out of their curriculum whenever possible.

The Trachtenberg system, which has been thoroughly tested in Switzerland, starts at the real beginning—in basic arithmetic where the student first encounters difficulties and begins to acquire an emotional attitude that will cripple him in all his mathematical work.

The ability to do basic arithmetic with the spectacular ease which the Trachtenberg system imparts, erases the fear and timidity that so hinder the student when faced with the impressive symbolism, the absoluteness of mathematical rigor. It is this emotional road-block, not inability to learn, that is the real reason why so many students hate mathematics, say the experts.

That short cuts make arithmetic easier to grasp and more palatable was proved conclusively by the armed forces during the last war. Bombardiers and navigators taking refresher courses in higher mathematics were able to cram several years' work into a few months when it had been simplified.

In Zurich, medical students, architects, and engineers find that the Trachtenberg system of simplified mathematics, enables them to pass the strict examinations necessary to complete their training. One of Switzerland's leading architects was enabled to continue with his chosen career only after attending the Institute where he learned the Trachtenberg method.

In Switzerland when people speak of the Mathematical Institute, they refer to it as the "School for Genius."

In an impressive test recently held in Zurich, students of the Trachtenberg system were pitted against mechanical brains. For a full hour the examiner called out the problems —intricate division, huge additions, complicated squarings and root findings, enormous multiplications.

As the machines began their clattering replies, the teen-age students quickly put down the answers without any intermediate steps.

The students beat the machines!

The students who proved as accurate as and speedier than the machines were not geniuses. It was the system—short and direct—which gave them their speed.

But it is not only in specialized professions that a knowledge of arithmetic is necessary. Today, in normal everyday living, mathematics plays an increasingly vital role. This is particularly true in America where we live in a welter of numbers. Daily the average man and woman encounters situations that require the use of figures—credit transactions, the checking of monthly bills, bank notes, stock market quotations, canasta and bridge and billiards scores, discount interest, lotteries, the counting of calories, foreign exchange, figuring the betting odds on a likely-looking steed in the fourth race, determining the chances of getting a flush or turning up a seven. And income taxes, among other blessings, have brought the need for simple arithmetic into every home.

The Trachtenberg system, once learned, can take the drudgery out of the arithmetic that is part of your daily stint.

The Swiss, noted for their business acumen, recognizing the brilliance and infallibility of the Trachtenberg system, today use it in all their banks, in most large business firms, and in their tax department. Mathematical experts believe that within the next decade the Trachtenberg system will have as far-reaching an effect on education and science as the introduction of shorthand did on business.

Published in book form for the first time, this is the original and authoritative Trachtenberg system. As you go through the book you will note that Professor Trachtenberg incorporated into his system a few points that were not original with him. These are on matters of secondary importance and are used for the purpose of greater simplification. To keep the record straight, we call attention to these points when they occur in the text.

The authors believe that anyone learning the rules put forth here can become proficient in the use of the Trachtenberg system.

The Trachtenberg Speed System of Basic Mathematics

CHAPTER ONE

Tables or no tables?

BASIC MULTIPLICATION

The aims of the Trachtenberg system have been discussed in the foreword. Now let us look at the method itself. The first item on the agenda is a new way to do basic multiplication: we shall multiply without using any memorized multiplication tables. Does this sound impossible? It is not only possible, it is easy.

A word of explanation, though: we are not saying that we disapprove of using tables. Most people know the tables pretty well; in fact, perfectly, except for a few doubtful spots. Eight times seven, or six times nine are a little uncertain to many of us, but the smaller numbers like four times five are at the command of everyone. We approve of using this hard-won knowledge. What we intend to do now is consolidate it. Later in this chapter we shall come back to this point. Now we wish to do some multiplying without using the multiplication tables.

Let us look at the case of multiplying by eleven. For the sake of convenience in explaining it, we first state the method in the form of rules:

MULTIPLICATION BY ELEVEN

1. **The last number of the multiplicand (number multiplied) is put down as the right-hand figure of the answer.**
2. **Each successive number of the multiplicand is added to its neighbor at the right.**
3. **The first number of the multiplicand becomes the left-hand number of the answer. This is the last step.**

In the Trachtenberg system you put down the answer one figure at a time, right to left, just as you do in the system you now use. Take an easy example, 633 times 11:

$$\underline{6\ 3\ 3} \quad \times \quad 1\ 1$$

answer will be here

The answer will appear under the 633, one figure at a time, from right to left, as we apply the rules. This will be our form for setting up the work from now on. The asterisks above the multiplicand of our example will quickly identify the numbers being used in each step of our calculation. Let us apply the rules:

First Rule

Put down the last figure of 633 as the right-hand figure of the answer:

$$\underline{6\ 3\ \overset{*}{3}} \quad \times \quad 1\ 1$$
$$3$$

Second Rule

Each successive figure of 633 is added to its right-hand neighbor. 3 plus 3 is 6:

$$\underline{6\ \overset{*}{3}\ \overset{*}{3}} \quad \times \quad 1\ 1$$
$$6\ 3$$

$$\underline{\overset{*}{6}\ \overset{*}{3}\ 3} \quad \times \quad 1\ 1$$

Apply the rule again, 6 plus 3 is 9:

$$9\ 6\ 3$$

Third Rule

The first figure of 633, the 6, becomes the left-hand figure of the answer:

```
 *
 6 3 3   ×  1 1
-------
6 9 6 3
```

The answer is 6,963.

Longer numbers are handled in the same way. The second rule, "each successive number of the multiplicand is added to its neighbor at the right," was used twice in the example above; in longer numbers it may be used many times. Take the case of 721,324 times 11:

```
7 2 1 3 2 4   ×  1 1
-----------
```

First Rule

The last figure of 721,324 is put down as the right-hand figure of the answer:

```
          *
7 2 1 3 2 4   ×  1 1
-----------
          4
```

Second Rule

Each successive figure of 721,324 is added to its right-hand neighbor:

```
        * *
7 2 1 3 2 4   ×  1 1
-----------
        6 4          2 plus 4 is 6
```

```
      * *
7 2 1 3 2 4   ×  1 1
-----------
      5 6 4          3 plus 2 is 5
```

```
    * *
7 2 1 3 2 4   ×  1 1
-----------
    4 5 6 4          1 plus 3 is 4
```

```
  * *
 7 2 1 3 2 4   ×   1 1
 ___________
   3 4 5 6 4         2 plus 1 is 3
```

```
 * *
 7 2 1 3 2 4   ×   1 1
 ___________
 9 3 4 5 6 4         7 plus 2 is 9
```

Third Rule

The first figure of 721,324 becomes the left-hand figure of the answer:

```
   *
   7 2 1 3 2 4   ×   1 1
 _____________
 7 9 3 4 5 6 4
```

The answer is 7,934,564.

As you see, each figure of the long number is used twice. Once it is used as a "number," and then, at the next step, it is used as a "neighbor." In the example just above, the figure 1 (in the multiplicand) was a "number" when it gave the 4 of the answer, but it was a "neighbor" when it contributed to the 3 of the answer at the next step:

```
     * *                      * *
 7 2 1 3 2 4   ×   11     7 2 1 3 2 4   ×   1 1
 ___________              ___________
     4                        3
```

Instead of the three rules, we can use just one if we apply it in a natural, common-sense manner, the one being "add the neighbor." We must first write a zero in front of the given number, or at least imagine a zero there. Then we apply the idea of adding the neighbor to every figure of the given number in turn:

```
      *
0 6 3 3   ×   1 1
------- 
      3 —there is no neighbor, so we add nothing!

0 6 3 3   ×   1 1
-------
  9 6 3 —as we did before

*   *
0 6 3 3   ×   1 1
-------
6 9 6 3 —zero plus 6 is 6
```

This example shows why we need the zero in front of the multiplicand. It is to remind us not to stop too soon. Without the zero in front, we might have neglected to write the last 6, and we might then have thought that the answer was only 963. The answer is longer than the given number by one digit, and the zero in front takes care of that.

Try one yourself: 441,362 times 11. Write it in the proper form:

```
0 4 4 1 3 6 2   ×   1 1
-------------
```

If you started with the 2, which is the right place to start, and worked back to the left, adding the neighbor each time, you must have ended with the right answer: 4,854,982.

Sometimes you will add a number and its neighbor and get something in two figures, like 5 and 8 giving 13. In that case you write the 3 and "carry" the 1, as you are accustomed to doing anyway. But you will find that in the Trachtenberg method you will never need to carry large numbers. If there is anything to carry it will be only a 1, or in later cases perhaps a 2. This makes a tremendous difference when we are doing complicated problems.

It is sufficient to put a dot for the carried 1, or a double dot for the rarer 2:

```
0 1 7 5 4   ×  1 1
1 9 ·2 9 4 —the ·2 is 12, from 7 plus 5
```

Try this one yourself: 715,624 times 11. Write it out:

```
0 7 1 5 6 2 4   ×  1 1
```

There is a 1 to carry under the 5 of the long number.

The correct answer to this problem is 7,871,864.

In the very special case of long numbers beginning with 9 followed by another large figure, say 8, as in 98,834, we may get a 10 at the last step. For example:

```
    9 8 8 3 4   ×  1 1
1 0 ·8 ·7 ·1 7 4
```

MULTIPLICATION BY TWELVE

To multiply any number by 12, you do this:

Double each number in turn and add its neighbor.

This is the same as multiplying by 11 except that now we double the "number" before we add its "neighbor." If we wish to multiply 413 by 12, it goes like this:

```
                  *
First step:  0 4 1 3  ×  1 2
                  6        double the right-hand figure and
                           carry it down
                           (There is no neighbor)
```

```
                    *  *
Second step:   0 4 1 3   ×   1 2
                   5 6       double the 1 and add 3

                  *  *
Third step:    0 4 1 3   ×   1 2
                 9 5 6       double the 4, add the 1

               *  *
Last step:     0 4 1 3   ×   1 2
               4 9 5 6       zero doubled is zero; add the 4
```

The answer is 4,956. If you go through it yourself you will find that the calculation goes very fast and is very easy.

Try one yourself: 63,247 times 12. Write it out with the figures spaced apart, and put each figure of the answer directly under the figure of the 63,247 that it came from. This is not only a good habit because of neatness, it also is worth its weight in gold as a protection against errors. In the particular case of Trachtenberg multiplication, we mention it because it will identify the "number" and the "neighbor." The next blank space in the answer, where the next figure of the answer will go, is directly below the "number" (in this example the figure that you must double). The figure to its right is the "neighbor" that must be added. The example works out in this way:

```
          *
0 6 3 2 4 7   ×   1 2
         ·4       double 7, 14; carry 1

        *  *
0 6 3 2 4 7   ×   1 2
      ·6 ·4       double 4, plus 7, plus 1 is 16; carry 1

      *  *
0 6 3 2 4 7   ×   1 2
    9 ·6 ·4       double 2, plus 4, plus 1 is 9
```

until you end up with:

```
0 6 3 2 4 7  ×  1 2
-----------
7 ·5 8 9 ·6 ·4
```

MULTIPLICATION BY FIVE, BY SIX, AND BY SEVEN

All these multiplications—5, 6, and 7—make use of the idea of "half" a digit. We put "half" in quotation marks because it is a simplified half. We take half the easy way, by throwing away fractions if there are any. To take "half" of 5, we say 2. It is really 2½, but we won't use the fractions. So "half" of 3 is 1, and "half" of 1 is zero. Of course "half" of 4 is still 2, and so with all even numbers.

This step is to be done *instantly*. We do not say to ourselves "half of 4 is 2" or anything like that. We *look* at 4 and *say* 2. Try doing that now, on these digits:

2, 6, 4, 5, 8, 7, 2, 9, 4, 3, 0, 7, 6, 8, 5, 9, 3, 6, 1

The odd digits, 1, 3, 5, 7, and 9, have this special feature of dropping the fractions. The even digits, 0, 2, 4, 6, and 8, give the usual result anyway.

MULTIPLICATION BY SIX

Now let us try out this business of "half." Part of the rule for multiplying by 6 is:

To each number add "half" of the neighbor.

Let us assume for the moment that this is all we need to know about multiplying by 6 and work out this problem:

```
0 6 2 2 0 8 4   ×   6
-------------
```

First step: 4 is the first "number" of the long number, and it has no neighbor so there is nothing to add:

```
            *
0 6 2 2 0 8 4   ×   6
-------------
            4
```

Second step: the second number is the 8, and its neighbor is the 4, so we take the 8 and add half the 4 (2), and we get 10:

```
          * *
0 6 2 2 0 8 4   ×   6
-------------
         •0 4
```

Third step: the next number is the zero. We add to it half its neighbor, the 8. Zero plus 4 is 4, and add the carry (1):

```
        * *
0 6 2 2 0 8 4   ×   6
-------------
       5 •0 4
```

Repeat this last step with the 2, the 2, the 6, and the zero, in turn:

```
0 6 2 2 0 8 4   ×   6
-------------
3 7 3 2 5 •0 4
```

Would you like to see how easy it is? Try it yourself on these two multiplications:

```
    0 4 4 0 4   ×   6
    ---------

0 2 8 6 8 8 4 2 4   ×   6
-----------------
```

The answer to the first problem is 26,424. The answer to the second one is 172,130,544.

What we have done gave the correct answer in these problems. However, it was not quite the full rule for multiplying by 6. The full rule is:

To each "number" add half the neighbor;
plus 5 if "number" is odd.

"If odd" means if the "number" is odd, it makes no difference whether the "neighbor" is odd. We look at the "number" and see whether it is odd or even. If it is even we merely add to it half of the neighbor. If it is odd, we add 5 to it and "half" of the neighbor, as we did just above. For instance:

```
0 4 4 3 0 5 2  ×  6
-------------
```

The figures 3 and 5, are odd. We see that as soon as we look at the multiplicand. When we come to work on the 3 and the 5 we shall have to add an extra 5, simply because of their oddness. It goes like this:

First step:

```
            *
0 4 4 3 0 5 2  ×  6
-------------
            2         2 is even and has no
                      neighbor; carry it down
```

Second step:

```
          * *
0 4 4 3 0 5 2  ×  6
-------------
         •1 2         Five is odd! 5 plus 5, plus
                      "half" of 2, is 11
```

Third step:

```
        * *
0 4 4 3 0 5 2  ×  6
-------------
        3•1 2         "half" of 5 is 2; then add
                      the carry
```

	* * (over 3 0)		
Fourth step:	0 4 4 3 0 5 2	× 6	
	8 3 ·1 2		*3 is odd! 3 plus 5 is 8*
	* * (over 4 3)		
Fifth step:	0 4 4 3 0 5 2	× 6	
	5 8 3 ·1 2		*4 plus "half" of 3*
	* * (over 4 4)		
Sixth step:	0 4 4 3 0 5 2	× 6	
	6 5 8 3 ·1 2		*4 plus "half" of 4*
	* * (over 0 4)		
Last step:	0 4 4 3 0 5 2	× 6	
	2 6 5 8 3 ·1 2		*zero plus "half" of 4*

The answer is 2,658,312. Of course, all this explanation is only for the sake of the greatest possible clarity in showing the method for the first time. In actual practice, it goes fast because the step of adding half the neighbor is a very simple one. With only a reasonable amount of practice it becomes more automatic than conscious.

You will see this more clearly, perhaps, if you go through these two for yourself:

0 8 2 3 4 × 6

0 6 2 5 0 1 8 8 × 6

The answer to the first problem is 49,404. The answer to the second one is 37,501,128.

The numbers that we have multiplied by 6 were long numbers. Would the method still work if we tried to multiply single digits, such as 8 times 6? Yes, it would, and, in fact, no change at all is needed. Try 8 times 6, using the same procedure:

$\underline{0\ 8} \times 6$

8 *there is no neighbor; 8 plus "half" the neighbor is 8*

$\underline{0\ 8} \times 6$

$4\ 8$ *zero plus "half" of 8 is 4*

When the number being multiplied is an odd digit, like 7, we must add the 5 at the first step. Of course, we do not add it at the second step, because zero is considered to be an even number:

$\underline{0\ 7} \times 6$

${}^{\bullet}2$ *7 plus 5, plus "half" of nothing*

$\underline{0\ 7} \times 6$

$4\ {}^{\bullet}2$ *zero plus "half" of 7, plus the carried 1*

Most people, probably, feel that they know the multiplication table for six by heart. More than half of the non-mathematical people, perhaps, have a feeling of confidence about it, even in cases where such confidence is not justified. That is not the point here. The techniques used in this multiplication method are going to be used again later in somewhat more complicated situations, and they will then be needed aside from any memorized tables. The best way to bring in these new processes is to do it on relatively familiar material. That is what we are doing now.

Also (and this is more important than it may sound), it is the way to start off on proper mental habits of calculation. We have all heard criticism of the reading habits of the average man, and about clinics which develop fast reading abilities. The critics say that too many people have the habit of reading letter by letter, spelling out whatever they are reading, or at least doing so to a great degree. We are urged to develop the habit of reading by recognizing a whole word or phrase at a time. Other points are brought up, too. They all amount to this: most people read badly to the extent that they have inefficient habits of reading.

Something along the same general line is true in arithmetic. A person has fallen into certain bad habits in the way that he goes about doing arithmetic, and the result is that he wastes some of his time and his energy. Only those, such as accountants, who spend most of their time working with figures eventually work out the proper procedures for themselves. The rest of us, even though we may not be occupied with calculations as our livelihood, can still learn these methods with a little effort and practice. Some of this material is indicated in this chapter and the next.

One of these mental steps, a very simple one, was mentioned already, when we were talking about using "half" the neighbor. We did a little practice at looking at a single figure, like 2 or 8, and saying immediately 1 or 4, without going through any mental steps. The answer should come into the mind as soon as we see the 2 or the 8, as though it were a reflex action. The reader would do well to go back to the figures offered for exercise and do them again.

Another of the correct mental steps is saying to ourselves only the result of adding the neighbor, or half the neighbor, like this:

$$\begin{array}{cccccc} & & * & * & & \\ 0 & 2 & 6 & 4 & \times & 6 \\ \hline & & 8 & 4 & & \end{array}$$

The 8 is the 6, plus half the 4. But do not say "half of 4 is 2, and 6 and 2 is 8." Instead, look at the 6 and the 4, see that half of 4 is 2, and say to yourself "6, 8." At first this will be difficult, so it may be better to say to yourself "6, 2, 8."

Another point that needs practice is the step of adding the 5 when the number (not the neighbor) is odd. Take this case:

$$\begin{array}{r} 0\ 6\ \overset{*}{3}\ \overset{*}{4} \\ \hline {}^{\bullet}0\ 4 \end{array} \quad \times \quad 6$$

The zero is the zero of 10, as the dot shows, and the 10 is 3 plus the 5 (because 3 is odd) plus 2 (half of 4). The correct procedure, at first, is to say "5, 8, 2, 10." After some practice this way, it should eventually be cut down to "8, 10." The 5 that comes in because 3 is odd should be added first, otherwise we may forget to add it.

In the same way, when there is a dot for a carried 1, this should be added before we add the neighbor (for times 11) or half the neighbor (for times 6). If we try to leave the carried 1 till after we add the neighbor we will sometimes forget it. In the example just above, the next figure of the answer is found like this:

$$\begin{array}{r} 0\ \overset{*}{6}\ \overset{*}{3}\ 4 \\ \hline 8\ {}^{\bullet}0\ 4 \end{array} \quad \times \quad 6$$

We look at the 6 and say "7," adding the dot; then we say "8," adding "half" the 3. At first it is better to look at the 6 and say "7," adding the dot, then say "1" for "half" of 3, then "8," and we write the 8.

When there is a dot and also a 5 to be added (because of oddness), say "6" instead of "5" and then add the number itself. This cuts out a step and is easy to get used to.

Take a pencil and try to use only the correct mental steps as you do these examples—the answers follow:

Times Eleven (add the neighbor):

1. 0 4 2 3 2 **2.** 0 4 7 4 9 2

Times Twelve (double and add the neighbor):

3. 0 4 2 3 2

4. 0 4 7 4 9 2

Times Six (add 5, if odd, and half the neighbor):

5. 0 2 2 2 2

6. 0 2 0 0 4

7. 0 4 2 3 2

8. 0 4 7 4 8

9. 0 2 9 0 6

10. 0 5 2 4 4

11. 0 3 8 6 5

12. 0 4 1 1 1

The answers are:

1. 46,552
2. 522,412
3. 50,784
4. 569,904
5. 13,332
6. 12,024
7. 25,392
8. 28,488
9. 17,436
10. 31,464
11. 23,190
12. 24,666

MULTIPLICATION BY SEVEN

The rule for multiplying by seven is very much like that for six:

Double the number and add half the neighbor; add 5 if the number is odd.

Suppose we wish to multiply 4,242 by 7. There are no odd digits in this number, so we shall not need to add the extra 5. For this example, the work goes the same as for six, except that now we double:

First step: 0 4 2 4 2* × 7

4 — *double the "number" 2*

Second step: 0 4 2 4* 2* × 7

9 4 — *double 4 plus half the neighbor*

Third step: 0 4 2* 4* 2 × 7 — 6 9 4 — *double 2 plus half the neighbor*

Fourth step: 0 4* 2* 4 2 × 7 — 9 6 9 4

Last step: 0* 4* 2 4 2 × 7 — 2 9 6 9 4 — *doubling zero we still have zero, but add half of neighbor*

Here is an example containing odd digits. The 3 and 1 are both odd:

First step: 0 3 4 1 2* × 7 — 4 — *double 2; there is no neighbor*

Second step: 0 3 4 1* 2* × 7 — 8 4 — *double 1, plus 5—because 1 is odd—is 7, and add half of 2*

Third step: 0 3 4* 1* 2 × 7 — 8 8 4 — *not odd: double 4 and add half of 1*

Fourth step: 0 3* 4* 1 2 × 7 — ·3 8 8 4

Last step: 0* 3* 4 1 2 × 7 — 2 ·3 8 8 4 — *doubling zero is zero, but add half of 3, plus the dot*

The correct mental steps are:

(1) Say "1" for the dot, if there is a carried 1.

(2) Look at the next number to work on and notice whether it is odd. If it is, add 5 to the carried 1, saying "6," or say "5" if there was no dot.

(3) Looking at the number and doubling it mentally, we say the sum of the 5 and this doubled figure. If the figure is 3, for instance, we say "5," then we say "11," because doubling the 3 to get 6 and adding it to the 5 can be done in one step.

(4) Looking at the neighbor, say 6 for example, we add half of it to what we have already. We were saying just now that we had 11. If the neighbor is 6, we next say "14."

Let us take this process a little at a time. The mental training involved in doing this sort of thing is very valuable because it develops the ability to concentrate, and concentration is practically the whole secret of success. It cannot be developed all at a time, however, and we can help ourselves by using several distinct stages in the following way:

First: Look at each of the following figures and immediately, without going through any intermediate steps, say aloud twice that number (looking at 3, say instantly "6" without saying "3" at all):

2, 4, 1, 6, 0, 3, 5, 1, 4, 3, 8, 2, 6, 3,
7, 5, 9, 2, 1, 0, 6, 3, 5, 2, 6, 8, 7, 4

Second: In each of the following pairs of numbers, look at the left-hand figure and say aloud its double (look at 3 and say "6"), then add its neighbor (for the pair 3 4 say "6, 10"). This is the fast way to multiply by 12:

2 1	3 4	2 0	1 1	2 2	0 2
2 7	1 5	6 0	7 1	4 5	0 9
3 2	3 8	7 4	5 2	8 2	4 1

Third: In each of the following pairs of numbers, look at the left-hand figure and say aloud its double, then add half its neighbor (look at 2 6, say "4, 7." This is "times 7" for even numbers:

2 6	2 7	4 0	6 1	2 6	4 4
0 4	2 2	2 9	8 1	8 8	8 9
6 6	4 3	6 7	4 9	8 1	0 7

Fourth: For each of the following numbers, look at the number and say "5," then say 5 plus the double of the number (looking at 3, say "5, 11. "):

7, 5, 3, 1, 9, 3, 7, 5, 1

Now go through it again!

Fifth: In each of the following pairs of numbers, look at the left-hand number, say "5," then say 5 plus the double of the number as we just did, then immediately add half the neighbor and say the result of adding this half (for 3 4, say "5, 11, 13"); this is times 7 for odd numbers:

1 0	1 2	1 6	1 8

(answers are 7, 8, 10, 11)

3 0	3 2	3 8	3 4
5 0	5 6	7 0	7 2

Now see how fast you can multiply by 7. Try first these numbers which are all even, so that there are no 5's to add, you only double the number and add half the neighbor:

0 2 0 2	0 2 2 2	0 6 0 2
0 4 4 4	0 6 4 2	0 8 4 6

Then we finish up with numbers containing some odd digits which have to have 5's added in:

0 2 2 3   0 3 0 2   0 2 5 4
(answers are 1,561, 2,114, and 1,778)

0 2 7 4   0 6 1 8   0 1 3 4

MULTIPLICATION BY FIVE

The rule for multiplying by 5 is like those for 6 and for 7, but simpler. Instead of adding in the "number" as we do for 6, or doubling it as we do for 7, we use the "number" only to look at. We look at it and see whether it is odd or even. If it is odd we add in the 5, as before:

Half the neighbor,
plus 5 if the number is odd.

Suppose we wish to multiply 426 by 5:

$$\underline{0\ 4\ 2\ \overset{*}{6}} \times 5$$

0   *look at 6, it is even; there is no 5 to add; there is no neighbor*

$$\underline{0\ 4\ \overset{*}{2}\ \overset{*}{6}} \times 5$$

3 0   *look at 2, it is even; use half of 6*

$$\underline{0\ \overset{*}{4}\ \overset{*}{2}\ 6} \times 5$$

1 3 0   *look at 4, it is even; use half of 2*

$$\underline{\overset{*}{0}\ \overset{*}{4}\ 2\ 6} \times 5$$

2 1 3 0   *look at zero, it is even; use half of 4*

Now if we had an odd digit in the multiplicand we would add 5:

```
      *
0 4 3 6   ×   5
-------
      0       as above

    * *
0 4 3 6   ×   5
-------
    8 0       the 3 is odd; say 5 plus 3

0 4 3 6   ×   5
-------
2 1 8 0
```

This is easy to do. There is very little figuring involved. It seems rather odd at first, though, because of the little mental twist that you have to do: you use the neighbor, rather than the digit at the place where you are working. Actually it is good practice in keeping one's place. Later on, in multiplying one long number by another long number, we shall find that a certain amount of concentration is needed to remember where we are in the number we are multiplying. This multiplication method for 5 is a little preliminary practice.

Try these, multiplying by 5 as just described:

1. 0 4 4 4
2. 0 4 2 8
3. 0 4 2 4 8 8 2
4. 0 4 3 4
5. 6 4 7
6. 0 2 5 6 4 1 3
7. 0 1 4 2 8 5 7

The answers are:

1. 2,220
2. 2,140
3. 2,124,410
4. 2,170
5. 3,235
6. 1,282,065
7. 714,285

MULTIPLICATION BY EIGHT AND NINE

For multiplying by 8 and 9 we have a new mental step, which gives further mental training. The new step consists of subtracting the "number" from 9 or from 10. Suppose we wish to multiply 4,567 by either 8 or 9; in both cases, the first step will be to subtract the right-hand figure of the long number (the 7) from 10. We begin by looking at the right-hand end of the 4,567 and saying "3." We should not go through the step of saying "7 from 10 is 3," it should be an immediate reaction. We look at the 7 and say "3." See how fast your reactions are—look at each of the following digits and call out instantly the result of taking it from 10:

7, 6, 9, 2, 8, 1, 7, 4, 2, 3, 9, 6, 5, 3, 1, 9

Part of the time we shall need to take the "number" from 9, instead of 10. In that case we would look at 7, for instance, and say instantly "2." Try this on the following digits, as fast as you can:

7, 8, 2, 4, 9, 5, 1, 7, 2, 0, 3, 8, 6, 5, 1, 0

Now you can multiply easily and rapidly by 9 without using the multiplication tables. The best way to make it clear is to state a rule, which you will not need to memorize because it will fix itself in your mind by a little practice. The rule would go like this:

Multiplication By Nine

1. **Subtract the right-hand figure of the long number from ten. This gives the right-hand figure of the answer.**

2. **Taking each of the following figures in turn, up to the last one, subtract it from nine and add the neighbor.**

3. At the last step, when you are under the zero in front of the long number, subtract one from the neighbor and use that as the left-hand figure of the answer.

Of course, in all these steps it is understood that if there is a dot (a carried one), you have to add it in as usual.

Here is an illustration of how it works out: 8,769 times 9.

```
0 8 7 6 9   ×  9
7 8 9·2 1
```

First: Take the 9 of 8,769 from 10, and we have the 1 of the answer.

Second: Take the 6 from 9 (we have 3) and add the neighbor, 9; the result is 12, so we write a dot and 2.

Third: 7 from 9 is 2, the neighbor (6) makes it 8, and the dot makes it 9.

Fourth: 8 from 9 is 1, and the neighbor makes it 8.

Fifth: This is the last step; we are under the left-hand zero. So we reduce the left-hand figure of 8,769 by one, and 7 is the left-hand figure of the answer.

Try this one yourself: 8,888 times 9.

```
0 8 8 8 8   ×  9
```

It ends in a 2 because 8 from 10 is 2. There are no dots in this case, no carrying, and the left-hand figure is a 7, namely the left-hand 8 less 1. So the correct answer is 79,992.

Here are a few to practice on, at first easy ones, then harder ones; the answers follow:

1. 0 3 3 **2.** 0 9 8 6 5 4 **3.** 0 8 6 7 3 3

4. 0 6 2 6   **5.** 0 8 0 5   **6.** 0 7 7 5 4 9 6 5

Answers: **1.** 297   **2.** 887,886   **3.** 780,597
4. 5,634   **5.** 7,245   **6.** 69,794,685

Multiplication By Eight

1. **First figure: subtract from ten and *double*.**
2. **Middle figures: subtract from nine and *double* what you get, then add the neighbor.**
3. **Left-hand figure: subtract *two* from the left-hand figure of the long number.**

Multiplying by 8 is the same as multiplying by 9 except for the doubling, and except that at the last step you take 2, not 1, from the left-hand figure of the long number. It goes like this:

```
      *
0 7 8 9  ×  8
-------
      2
```

This 2 comes from taking the 9 from 10 and doubling. Then the 8 of 789 is one of the "middle number" type so we take the 8 from 9 and double and we add the neighbor:

```
    * *
0 7 8 9  ×  8
-------
   ·1 2
```

The 7 is also a "middle" number; we are not at the end until we reach the zero in front of 789. So for 7 we double 2 (7 from 9 is 2), and this 4 is added to the 8:

```
  * *
0 7 8 9  ×  8
-------
 ·3·1 2
```

Finally the left-hand 7 is reduced by 2 to give 5 (plus the carried dot, of course):

$$\begin{array}{llll} \overset{*}{0} & \overset{*}{7} & 8 & 9 \quad \times \quad 8 \\ \hline 6 & {}^{\bullet}3 & {}^{\bullet}1 & 2 \end{array}$$

Notice how much simpler and easier this really is, once the method has become familiar, than conventional multiplication. By the conventional method we must not only be sure of the multiplication tables (and many persons are unsure of the 8 times 7, 8 times 8, and 8 times 9 that we need here), we must also carry 7, then 7 again, giving rise to possible errors. In contrast, the non-table way requires carrying only a 1.

Of course, the method is not really mastered until it can be used without thinking of any "rules." A little practice makes the procedure sufficiently automatic to accomplish this. Half an hour, or an hour, to practice this method is actually very little practice indeed when we compare it to the hours of repeated drill that children in school devote to learning the multiplication tables.

Try these for practice, multiplying each number by 8:

0 7 3 (answer, 584)	0 4 9 (answer, 392)
0 6 9	0 9 8
0 7 7 7 7	0 8 5 8 6
0 6 2 8 8	0 3 6 6 9

The same process works equally well on single-figure numbers. Suppose we are multiplying by 9 (no doubling!) and we wish to multiply the number 7. There is no "middle" number, so we subtract the 7 from 10 as our usual first step, then we subtract 1 from the 7 as our usual last step. It looks like this:

$$\frac{0\ 7}{6\ 3} \times 9$$

These single-digit numbers form a simple pattern, like this:

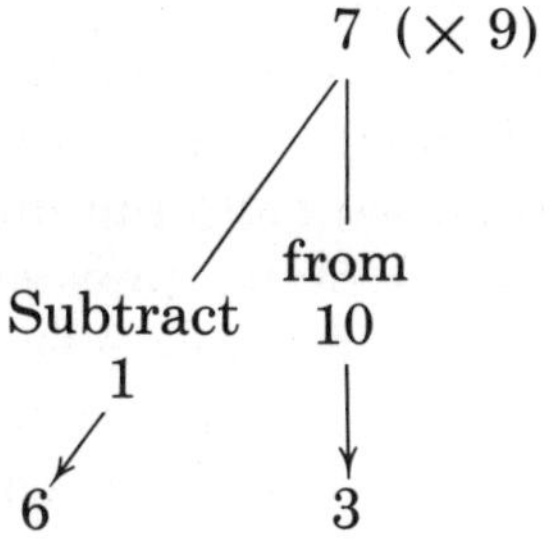

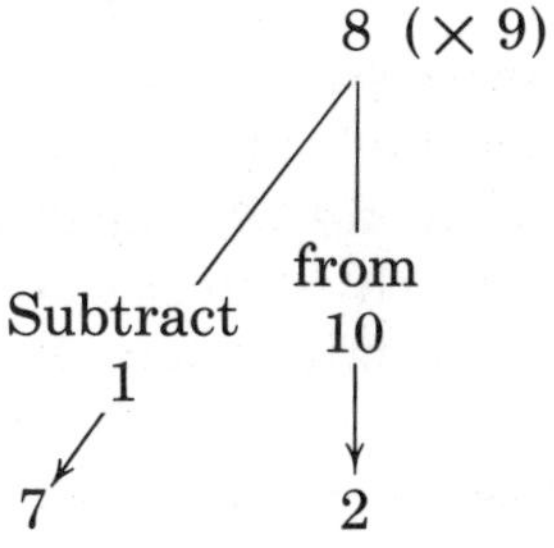

Similarly we can think of multiplying by 8, only it is "from 10" doubled and "subtract 2":

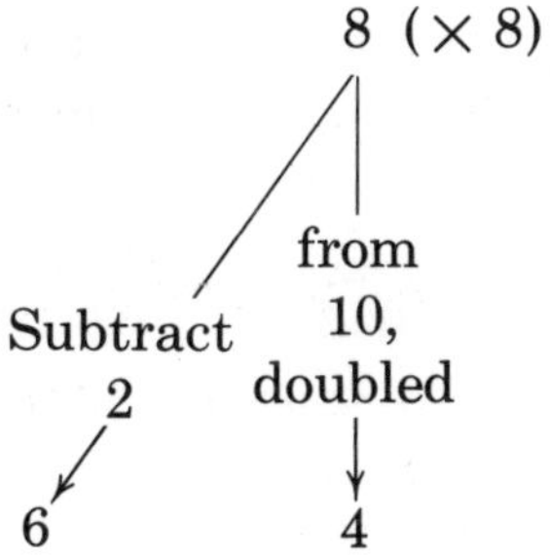

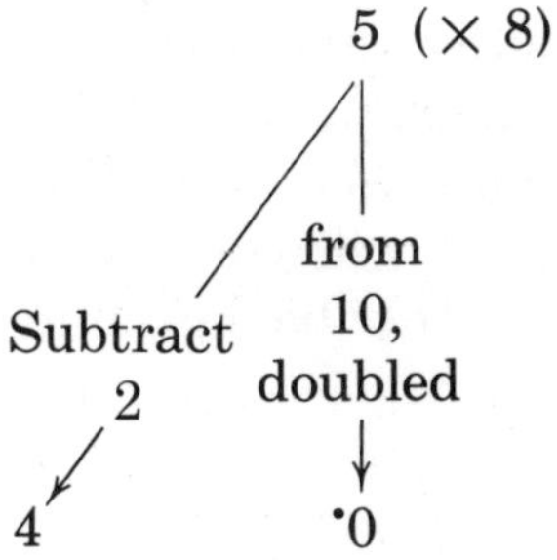

With 8 we would have to carry a 1 if we multiplied the smaller digits (1 to 5) this way. But this is not usually necessary because we all know the multiplication tables for these small numbers. It is only at the higher numbers (7 times 8), that people are likely to have a little trouble. For these, the diagram for 8 works without carrying. For 9 there is never anything to be carried.

Anyone who has trouble with the multiplication tables can use these diagrams to fix the uncertain parts in his memory. We need not go through the actual calculation each time. If one writes these diagrams out a few times, or even visualizes them, they will provide a background for the fact that 7 times 9 is 63, and so on. A background is all that is really needed. Only isolated facts are hard to remember. Suppose, for instance, that for several months you have not seen someone who used to be a close friend. You may not remember his telephone number, because telephone numbers are isolated facts. But you will probably remember the telephone exchange, because the exchanges have a pattern: they are assigned according to the district of the city where he lives, and you will remember approximately where he lives. Fitting a fact into any sort of pattern will fix it in the mind. In the case of mathematics, the best pattern is a derivation of the fact. A mathematician does not remember a theorem "cold" as a pure act of memory. Usually a sketchy idea of the proof or derivation of the theorem is attached to the idea of the theorem itself. That is what will happen if one practices with the diagrams. When 7 times 9 is called for, "63" will come to the forefront of the consciousness, and in the background the diagram will be present to prompt it.

MULTIPLICATION BY FOUR

Most people, even the least mathematical, feel confident of their ability to multiply by 4. For the sake of completeness, however, we show now how it could be done by a procedure similar to the ones we have been considering.

We do this by combining two ideas that we have already had. The first one is multiplication by 9, as we saw above, and the second one is taking "half" (the smaller half), and adding 5 if odd. To be precise, multiplying by 4 is the same as multiplying by 9 except for one thing: instead of "adding the neighbor," as we did with 9, we now add "half" the neighbor, plus 5 if the number is odd, as usual. Stated in full this means:

1. **Subtract the right-hand digit of the given number from ten, and add five if that digit is odd.**
2. **Subtract each digit of the given number in turn from nine, add five if the digit is odd, and add half the neighbor.**
3. **Under the zero in front of the given number, write half the neighbor of this zero, less 1.**

Example 1: 20,684 times 4

First step: Subtract the 4 of 20,684 from 10:

```
          *
0 2 0 6 8 4   ×   4
-----------
          6
```

Second step:

```
        * *
0 2 0 6 8 4   ×   4
-----------
        3 6
```

the 3 is 8 from 9 plus half of 4

Third step:

```
      * *
0 2 0 6 8 4   ×   4
-----------
      7 3 6
```

the 7 is 6 from 9 plus half of 8

Fourth step:

```
    * *
0 2 0 6 8 4   ×   4
-----------
   ·2 7 3 6
```

the ·2 is zero from 9 plus half of 6

Fifth step:

```
  * *
0 2 0 6 8 4  × 4
-----------
   8·2 7 3 6        the 8 is 2 from 9 plus the dot
```

Last step:

```
* *
0 2 0 6 8 4  × 4
-----------
0 8·2 7 3 6         the zero is half of 2 less 1
```

Example 2: We did not need to "add 5" in Example 1, because all the digits of 20,684 happen to be even. Here is a case where some are odd. Multiply 365,187 by 4.

First step:

```
            *
0 3 6 5 1 8 7  × 4
-------------
            8       7 from 10 is 3; then add
                    5 because 7 is odd
```

Second step:

```
          * *
0 3 6 5 1 8 7  × 4
-------------
          4 8       8 from 9 plus half of 7
```

Third step:

```
        * *
0 3 6 5 1 8 7  × 4
-------------
       ·7 4 8       the ·7 is 1 from 9 plus 5
                    plus half of 8
```

Fourth, Fifth, and Sixth steps: We repeat as before. Remember that 3 and 5 are odd, and call for the added 5:

```
0 3 6 5 1 8 7  × 4
-------------
 ·4 6·0·7 4 8
```

Last step:

```
0 3 6 5 1 8 7  × 4
-------------
1·4 6·0·7 4 8       the 1 is half of 3, less 1
                    plus the dot
```

For practice, if this is desired, we may apply these methods to the following examples:

1. 0 2 6 8 8 × 4 **3.** 0 2 4 7 8 4 7 × 4

2. 0 8 6 0 4 4 2 × 4 **4.** 0 5 4 6 1 8 × 4

Answers: **1.** 10,752 **2.** 3,441,768 **3.** 991,388 **4.** 218,472

Only a very little practice is required to make this become easy, as compared to the practice given to the multiplication tables. After a few hours, the operations come to us quite naturally.

MULTIPLICATION BY OTHER DIGITS

Multiplication By Three

Multiplying by 3 is similar to multiplying by 8 but with a few exceptions. Instead of adding the neighbor, as we did for 8, we now add only "half" the neighbor. It is understood, of course, that we also add the extra 5 if the number is odd. Adding half the neighbor always carries with it the extra 5 for odd numbers. Multiply 2,588 by 3.

1. **First figure: subtract from ten and *double*. Add five if the number is odd.**
2. **Middle figures: subtract the number from nine and double what you get, then add half the neighbor. Add five if the number is odd.**
3. **Left-hand figure: divide the left-hand figure of the long number in *half*; then subtract two.**

First step: 0 2 5 8 8* × 3

4 — *4 is 8 from 10, doubled; no neighbor*

Second step: 0 2 5 8* 8* × 3

6 4 — *6 is 8 from 9, doubled; plus half of 8*

Third step:

```
     * *
 0 2 5 8 8   ×   3
 ---------
     ·7 6 4        5 from 9, doubled; plus 5,
                   plus half of 8
```

Fourth step:

```
   * *
 0 2 5 8 8   ×   3
 ---------
   ·7 ·7 6 4
```

Last step:

```
 * *
 0 2 5 8 8   ×   3
 ---------
 0 ·7 ·7 6 4       zero is half of 2 plus the dot,
                   minus 2
```

At the last step we obtain the left-hand digit of the answer from the leftmost digit of the given number, as always. In multiplying by 8 we obtained this last figure of the answer by reducing the leftmost digit of the given number by 2. Now in multiplying by 3 we reduce half of that digit by 2. Sometimes, as in the example, half of the leftmost digit is only 1, and sometimes it is zero. In every such case it is also true that we have a carried 1 or a carried 2, so when we reduce by 2 we have zero. The example illustrated this.

Multiplication By Two

Multiplying by 2 is of course trivial. In terms of our procedures, we multiply each digit of the given number in turn by 2, and we do not use the neighbor. (We can double a number by adding it to itself, so even the times 2 table need not be memorized.)

Multiplication By One

Multiplying by 1 does not change a number. Any number, of any length, when multiplied by 1 remains itself:

17,205 times 1 is 17,205, for instance. Hence, the rule would be: copy down each digit of the given number in turn.

These last few rules, for multiplying by the small digits, have been included here mainly for the sake of completeness.

It is important to notice, however, that in all cases, for multiplication by any digits, the operations really required are few in number and all are simple. Subtracting from 9, doubling, taking "half," and adding the neighbor—these are the only operations involved. Practicing for an hour or two will make them seem natural and automatic.

That is why Professor Trachtenberg believed that the methods of this chapter would be particularly helpful to children. Long before they can memorize all the multiplication tables, they can be using these new methods with ease. Of course, this also enables them to perform any multiplication with numbers of any length. Each one-digit multiplication, by the rules above, gives a partial product in the conventional setup, and the total is found by adding columns as usual:

3 7 6 5 4 × 4 9 8	
3 0 1 2 3 2	*use the rule for multiplying 37,654 by 8*
3 3 8 8 8 6	*use the rule for 9*
1 5 0 6 1 6	*use the rule for 4*
1 8 7 5 1 6 9 2	*the answer, by adding columns*

Thus, a child who has only just learned to do the simplest kind of addition and subtraction can almost immediately perform long multiplications.

Note: Adults, also, are always taught the methods of this chapter in the Trachtenberg system. For adults, the purpose is different. They have already spent hundreds of hours, in their youth, memorizing the multiplication tables and know most of them very well. This new method fills the gaps. The psychological effect of looking at the matter from a new point of view is such that it fixes the doubtful parts of the tables firmly in their minds. Furthermore, as we mentioned previously, the novelty of this method awakens a fresh interest in the subject, and that in itself is half the battle. The experience of the Trachtenberg Institute over a thirteen-year period demonstrates the importance of these points.

SUMMARY

Rules will not be needed after a reasonable amount of practice. The act of working out examples makes the process become semi-automatic, and that is the best way to learn it. Nevertheless, for the convenience of any who may desire it, we repeat here the methods presented in this chapter. It is understood in the statement of these rules that the "number" is that digit of the multiplicand just above the place where the next digit of the answer will appear, and the "neighbor" is the digit immediately to the right of the "number." When there is no neighbor (at the right-hand end of the given number), the neighbor is zero—that is, it is ignored. Also, a zero is to be written in front of the multiplicand to remind us that a digit of the answer may appear there.

TO MULTIPLY BY	THE PROCEDURE IS THIS:
11	Add the neighbor.
12	Double the number and add the neighbor.
6	Add 5 to the number if the number is odd; add nothing if it is even. Add "half" the neighbor (dropping fractions, if any).

7 Double the number and add 5 if the number is odd, and add "half" the neighbor.

5 Use "half" the neighbor, plus 5 if the number is odd.

9 First step: subtract from 10.
Middle steps: subtract from 9 and add the neighbor.
Last step: reduce left-hand digit of multiplicand by 1.

8 First step: subtract from 10 and double.
Middle steps: subtract from 9, double, and add the neighbor.
Last step: reduce left-hand digit of multiplicand by 2.

4 First step: subtract from 10, and add 5 if the number is odd.
Middle steps: subtract from 9 and add "half" the neighbor, plus 5 if the number is odd.
Last step: take "half" the left-hand of the multiplicand and reduce by 1.

3 First step: subtract from 10 and double, and add 5 if the number is odd.
Middle steps: subtract from 9 and double, add 5 if the number is odd, and add "half" the neighbor.
Last step: take "half" the left-hand digit of the multiplicand and reduce by 2.

2 Double each digit of the multiplicand without using the neighbor at all.

1 Copy down the multiplicand unchanged.

0 Zero times any number at all is zero.

CHAPTER TWO

Rapid multiplication by the direct method

In Chapter One we saw how basic multiplication can be performed without the conventional multiplication tables. By means of these new ideas we have been able to refresh our knowledge of the tables and to clear up any uncertain spots that might have existed. We should now have more confidence in our ability to use the tables quickly and accurately whenever we wish to do so.

With this new approach to basic multiplication we have become accustomed to using a pair of digits in the multiplicand to give each figure of the answer. The "number," you will recall, is the digit just above the blank space where the next digit of the answer will appear; the "neighbor" is the digit of the multiplicand immediately to the right of the "number." Such number-neighbor pairs, in the position mentioned, will be used again in this chapter but with a variation.

We are now ready to take the next step in our method of condensing the multiplication process. We shall learn how to multiply any number by any other number no matter how long they are, and arrive at an immediate answer without any intermediate steps. The condensed form of multiplying 625 by 346, for instance, will look like this:

$$\frac{0\ 0\ 0\ 6\ 2\ 5}{2\ 1\ 6\ 2\ 5\ 0} \times 3\ 4\ 6$$

We are going to learn how to do multiplication in this form. Nothing else will be written down. The usual three rows of intermediate figures will not be used. We shall write down the problem, whatever it may be, and then we shall immediately write the answer.

We have two ways of accomplishing this. Each has its own advantages in certain situations, yet both of them are always capable of supplying the right answer. Fortunately, they have a great deal in common so it is easy to learn both. In the present chapter we are going to consider the one that we call the "direct" method of multiplication. It is most appropriate when the numbers to be multiplied contain small digits, like 1's, 2's, and 3's. Then in the next chapter we shall use the other method, which we call "speed multiplication." It consists of the direct method plus a new feature. This added feature takes care of the difficulties we meet in numbers containing large digits such as 987 times 688.

Either method can be used in any given problem. Both methods always give the right answer. Just now we mentioned a reason for preference, but it is entirely a matter of convenience, and in a particular case you would make the choice according to your personal judgment.

Incidentally, it may be mentioned that something like the direct method was probably used by mathematical speed performers before the Trachtenberg system was introduced. These "mathematical wizards," who astonished audiences by spectacular feats of mental calculation, were usually rather secretive about their techniques but it seems they must have used something similar to our direct method—perhaps with variations.

Let us begin with a simple example of the direct method and work up to more complicated cases, meaning that we shall first multiply a relatively small number by another relatively small number.

SHORT MULTIPLICANDS: TWO DIGITS BY TWO DIGITS

Suppose we wish to multiply 23 by 14. We write it out in this form:

$$\underline{0\ 0\ 2\ 3} \times 1\ 4$$

(*answer will be here*)

When multiplying by a two-digit multiplier, we always put two zeroes in front of the multiplicand, as shown above.

The answer will be written under the 0023, one digit at a time, beginning at the right. That is to say, we shall write the last digit of the answer under the 3, and fill in the rest of the answer one digit at a time to the left.

First step: Multiply the right-hand figure of the multiplicand, the 3 of 23, by the right-hand figure of the multiplier, the 4 of 14. For the answer we put down the 2 of 12 and carry the 1 (use a dot ˙):

$$\underline{0\ 0\ 2\ \overset{*}{3}} \times 1\ \overset{*}{4}$$

˙2 — *3 times 4 is 12; we write 2, carry 1*

Second step: We obtain the next figure of the answer, the one that is to go under the 2 of 23, by finding *two* numbers (two partial products) and adding them together. The first of these two is 8, which comes from 2 times 4:

$$\underline{0\ 0\ \overset{*}{2}\ 3} \times 1\ \overset{*}{4}$$
$$\cdot 2$$

The second of the two partial products is obtained by multiplying the other figures, 3 and 1:

$$\underline{0\ 0\ 2\ \overset{*}{3}} \times \overset{*}{1}\ 4$$
$$\cdot 2$$

Now add the two partial products together: 8 plus 3 is 11. This is what we want. But we must add the carried 1, so the next figure of the answer is 12; that is, a written 2 and a carried 1:

$$\underline{0\ 0\ 2\ 3} \times 1\ 4$$
$$\cdot 2\ \cdot 2$$

2 times 4 is 8; 3 times 1 is 3; 8 plus 3 is eleven; add the carry

Last step: Multiply the left-hand figure of the multiplicand, the 2 of 23, by the left-hand figure of the multiplier, the 1 of 14:

$$\underline{0\ 0\ \overset{*}{2}\ 3} \times \overset{*}{1}\ 4$$
$$3\ \cdot 2\ \cdot 2$$

2 times 1 is 2, and the carried 1 makes it 3

In this example we do not need to use the left-hand zero that we wrote in front. It is there to make room for a carried figure whenever there happens to be a figure of 10 or over. In this example we had only a 3.

The second step is the new one. We used two figures to get one figure of the answer. We added the partial product 8 and the partial product 3, and had 11 to use in our answer.

The 8 and the 3 came from multiplying two pairs of figures which we will call the "outside pair" and the "inside pair."

```
0 0 2 3   ×   1 4
    ·2        ↓ ↓
  ·2          3+8   3 plus 8 is 11,
                    plus carry
```

The rule for finding these pairs is this: the figure of the multiplicand that we are working on at the moment, (that is to say, the figure *just above* the next figure of the answer to be found,) is part of the "outside pair;" in the example above, it is the 2 of 23. The other figure of this "outside pair" is the right-hand figure of the multiplier, because that is the outside one; the 4 of 14 above. The other pair, the "inside pair" 3 and 1, is made up of the two figures immediately inside the pair of figures that you have just used; the 3 of 23 and the 1 of 14.

The "outside pair" and "inside pair" will be used frequently, so let us make clear what they are by these three examples:

46×87		72×34		28×92	
4 × 7	*outside pair*	7 × 4	*outside pair*	2 × 2	*outside pair*
6×8	*inside pair*	2×3	*inside pair*	8×9	*inside pair*

46×87   72×34   28×92

Take the case of 38 times 14:

0 0 3 8 × 1 4

First step: The first thing we do is multiply 8 by 4 and get 32. We put down the 2 and carry the 3.

```
       *      *
0 0 3  8  ×  1 4
-----------
      ···2
```

Second step: To find the next number we now use the outside and inside pairs. The figure of 38 that we are now working on is the 3, because the 3 is directly above the place where the next figure of our answer will appear. So this 3 is part of the outside pair. What is the other figure of the outside pair? It is one of the figures of 14, and obviously it has to be 4. That is the outside figure of the 14. The inside pair comes just inside those two (8 and 1):

```
0 0 3  8  ×  1 4
-----------
      ···2
```

Now multiply: 3 times 4 is 12, and 8 times 1 is 8. Add these two partial products, the 12 and the 8, and you have 20. A carried 3 must be added in, so the result is 23. Put down the 3 and carry the 2.

```
0 0 3  8  ×  1 4
-----------
   ··3 ···2
```

Last step: Multiply the two left-hand figures, the 3 of 38 and the 1 of 14. This gives 3. The carried 2 makes it 5:

```
    *         *
0 0 3  8  ×  1 4
-----------
0 5 ··3 ···2
```

Here are two examples worked out, but shown in short form. The underlined figures are carried figures. See if you can figure out for yourself where the numbers shown in the "work" line came from:

	0 0 3 2	× 2 2
answer:	7 ·0 4	
work:	6 6 4	
	+ +	
	1 4	

	0 0 6 6	× 3 4
answer:	2 2 ····4 ··4	
work:	18 24 4	
	+ +	
	4 18	
	+	
	2	

For your own satisfaction you may like to try one or two like these by your own efforts. Here are several, with the answers following:

1. 0 0 3 1	× 1 5	**4.** 0 0 3 4	× 2 1
2. 0 0 1 7	× 2 4	**5.** 0 0 4 2	× 2 6
3. 0 0 7 3	× 6 4	**6.** 0 0 4 8	× 5 2

Answers: **1.** 465 **2.** 408 **3.** 4,672 **4.** 714 **5.** 1,092 **6.** 2,496

If you reflect on what you have just been doing, you will see that it is a very natural procedure. You have been multiplying two two-digit numbers together. To do this, you have been finding the right-hand digit of the answer by multiplying the two right-hand digits, as with 23 times 14 you multiplied 3 by 4. To find the left-hand digit of the answer you multiplied the two left-hand digits, as with 23 times 14 you multiplied 2 by 1. In between, to get the middle figures of the answer, you used the outside and inside pairs. Each pair consists of two digits that are multiplied together, so each pair gives one number, and these two numbers are added to find part of the answer.

We are going to use these inner and outer pairs again in what follows. In fact, we shall use them to a greater extent. Of course, in doing a problem yourself you do not need to write out the pairs by drawing curved lines touching the digits. That was done in the early paragraphs only to explain what was meant. In actual work it is possible to identify the outside pair by the fact that it contains the digit of the multiplicand directly above the next blank space, that is, the place where the next figure of the answer will appear. The inside pair is the pair of figures just inside the two figures of the outside pair, as the curved lines showed in the diagram.

LONG MULTIPLICANDS

When the multiplicand is a long number, all we need to do is repeat the second step as many times as the long number requires. For instance, suppose you wish to multiply 312 by 14. This is only three digits instead of two, but it will illustrate the point:

First step: Multiply the right-hand digit of 312 by the right-hand digit of 14:

$$\begin{array}{c} \underline{0\ 0\ 3\ 1\ \overset{*}{2}} \quad \times \quad 1\ \overset{*}{4} \\ 8 \end{array}$$

Second step: Now use the outside and inside pairs. The next figure we work on is the 1 of 312. It is the figure directly above the place where the next digit of the answer will go. So the 1 of 312 is part of the outside pair:

```
          0 0 3 1 2  ×  1 4
answer:         6 8
work:           4          outside pair, 1 times 4, gives 4;
                +          inside pair, 2 times 1, gives 2;
                2          4 plus 2 is 6
```

Third step: This is the second step over again, except that we move the pairs. That is, we have different pairs of numbers. But it is still true that the next figure we use of 312, the one directly above the next space to fill in, is part of the outside pair. In this example, 3 is part of our new outside pair. So we have:

```
          0 0 3 1 2  ×  1 4
answer:       ·3 6 8
work:        12            outside pair, 3 times 4, gives 12;
              +            inside pair, 1 times 1, gives 1;
              1            12 plus 1 is 13; put down 3 and
                           carry 1
```

Last step: To find the left-hand figure of the answer, multiply the two left-hand figures together, 3 times 1, and add the carried 1:

```
              *             *
          0 0 3 1 2  ×  1 4
answer:     4·3 6 8
work:     3×1
           +
          dot
```

Later on we shall need to extend the curved lines over the zeroes in front of the multiplicand. Let us do it here, just to see how it works. Remember:

ANY NUMBER TIMES ZERO IS ALWAYS ZERO

In multiplying, zero "annihilates" every other number. A million times zero is zero. Use this fact, and do the last step as if it were a middle step:

```
Last step:  0 0 3 1 2   ×   1 4
            ---------
   answer:    4·3 6 8
            ---------
     work:    0
              +
              3
              +
             dot
```

The outside pair, zero times 4, gives zero. The inside pair, 3 times 1, gives 3. Zero plus 3 is 3. Adding the dot, we get 4, the same answer that we had before—it has to be the same because that is the right answer. This shows that we can do the last step by the same method as the second and third steps; that is, we can find it by using outside and inside pairs, *instead of* using a special rule for the last step.

Every time we show a line of figures and mark it "work," it is understood that in actual practice this work would be done mentally. We have shown the work explicitly for the purpose of explanation only. When you work a problem, you will do it by writing only the two figures that you wish to multiply and the answer.

The examples we had just now show again how the outside pair is placed. It is always determined by the fact that it contains the figure of the multiplicand which is immediately above the next figure of the answer to be found:

```
                 0 0 5 3 2 1 0 4 0 2   ×   6 4
                 -------------------
partial answer:            6 5 7 2 8
```

The other end of the line must go to the right-hand figure of the two-digit multiplier, because that is an "outside" digit. Then the inside pair is made up of the two figures just inside of the ones that we have used.

In actual use, you will probably find that it is a good idea to mark the positions of the figures in the outside and inside pairs by covering parts of the numbers with your fingers. This is really very little trouble, and it prevents errors that might occur from losing your place momentarily. In the case of three digits times two digits, like 312 times 14, there is not much danger of losing the place, but soon we shall be looking at much longer numbers. In any case, it is certainly advisable to write the figures clearly and spaced apart, and to write the answer directly under the figure it belongs to. Neatness is an aid in avoiding unnecessary errors. It applies not only to what we are doing here, but to conventional multiplication, to any kind of division, and to addition and subtraction. Neatness is a good habit to develop.

Here is a way to check yourself on how well you understand the method. Below is the completed example of the multiplication of 311 by 23. The answer is under the 311, and a line of work, supposedly mental, will be under the answer. Now cover both answer and work with a scrap of paper, and calculate the right-hand figure of the answer, mentally. Move the paper far enough to see the first figure of the answer, and you will know whether you are right. Then calculate mentally the next figure of the answer, and when you have it, move the paper just enough to expose this next figure and see whether you are right. If you are not, move the paper to expose the "work" line for that figure, and you will see where the figure came from. In this "work" the figures under each other *are to be added together* to get the figure of the answer:

	0	0	3	1	1	×	2 3
answer:		7	·1	5	3		
work:		0×3	3×3	1×3	1×3		
		3×2	1×2	1×2			
		(dot)					

THE ZEROES IN FRONT

In the examples we have seen so far, we had two zeroes in front of the number to be multiplied, but one would have been enough as it turned out. Sometimes, though, we need two zeroes in front. Consider this example:

	0	0	5	2	2	×	3 1
answer:	1	·6	·1	8	2		
work:	0×1	0×1	5×1	2×1	2×1		
	+	+	+	+			
	0×3	5×3	2×3	2×3			
	+	+					
	dot	dot					

This time we had a number under the second zero, the one all the way to the left. Notice, though, that the number is simply the dot (or carried 1). The figures of 31 gave nothing, because they were both annihilated by multiplying by zero. Only the carry remained.

That explains why we did not need two zeroes in our previous examples. One zero in front of the multiplicand was sufficient, because there was nothing to carry at the last step.

GENERAL RULE: When multiplying by a multiplier of any length, put as many zeroes before the multiplicand as there are digits in the multiplier.

Sometimes, as we saw, we don't need all the zeroes, but following the rules will never do any harm. If we go ahead and try to use two places when we need only one, we shall merely find that at the last step we have nothing to write.

So far we have considered only two- or three-digit multiplicands, but long numbers, like 241,304, are handled by the same method. We simply repeat the action of multiplying together two pairs of digits and adding the results. Suppose you wish to multiply 241,304 by 32:

```
           0 0 2 4 1 3  0 4   ×   3 2
answer:    ----------------
                   7 ·2 8
work:              6  0
                   0 12
                  dot
```

This is as far as we have gone previously, using three digits in the multiplicand. The next step is done in the same manner:

```
           0 0 2 4 1 3 0 4   ×   3 2
answer:    ---------------
                 ·1 7 2 8
work:              2            the 2 is 1 times 2; the 9 is
                   +            is 3 times 3
                   9
```

Of course, we add together the 2 and the 9, and this 11 is the next figure of the answer. We write 1 as part of the answer and carry 1 by writing a dot. Then we continue moving to the left. The next pairing is done by using 4,1 against 3,2, and it gives us 4 times 2 plus 1 times 3. The complete solution is this:

	0	0	2	4	1	3	0	4	×	3 2
answer:		7	·7	·2	·1	7	·2	8		
work:		0	4	8	2	6	0	8		
		6	12	3	9	0	12			
		dot	dot	dot		dot				

The zero farthest to the the left is not used. This is one of the cases where there is nothing to carry at the last step, so there will be nothing to write under this last zero. We showed it only for the sake of completeness. The rule calls for two zeroes when the multiplier has two figures. One zero is wasted here, but after all, what does a zero cost? *Nothing!*

Here is a little "thought-problem": You should be able to answer it without doing any direct calculation. Now that you know, from one of our examples, that 311 times 23 is 7,153, what is 31,100 times 23? The only difference is that now we have two zeroes at the end of the multiplicand. What is the answer now? Decide before you read the next paragraph.

The answer, as you no doubt said to yourself, is 715,300. The extra two zeroes at the end of the multiplicand have come over into the answer. *This is always true.* Any zeroes that may be tagged on at the end of any multiplicand—no matter how many zeroes or what the multiplicand may be—must be carried down directly into the answer, at the end of the answer.

Which of the four possible ways did you use to decide on this? Perhaps you were clever enough to use more than one of them at the same time. In any case, here are the four ways:

(*1*) *Guess-work.* This is a rather rude word, so you may prefer one of the synonyms. Non-mathematicians usually refer to this as "common-sense." Mathematicians refer to it as "mathematical intuition." Under either name it frequently gives the wrong answer, but often enough it works.

(*2*) *Memory.* From your school days you may have remembered how such things should go. If the memory was vague it counts as half memory and half "common-sense."

(*3*) *Multiplication by zero.* We know that zero times any number is still zero. Obviously, when we start multiplying the 23 into the two zeroes at the end of 31,100, we keep on getting zeroes until we hit the right-hand 1 of 31,100. Adding two zeroes is still zero. We get nothing different from zero until we start in on the 311 part of 31,100, and then we get the same thing as 311 times 23.

(*4*) *Rearrangement of factors.* This is the method that a mathematician would use. The basic idea is that when we multiply together more than two numbers, it does not make any difference how we group them, as long as we use them all eventually in multiplying. For instance, take 2 times 3 times 4. Doing it the straightforward way, we would say 2 times 3 is 6, then 6 times 4 is 24, so the result is 24. But we could, if we wished, begin by multiplying the 3 by the 4: 2 times 3 times 4 is 2 times 12, which is again 24. Further, we could rearrange them: 2 times 3 times 4 is the same as 2 times 4 times 3, or 8 times 3. Again the result is 24.

Now apply this to the case of 31,100 times 23. We think of it as 311 times 100 times 23. Rearrange the numbers: this is the same as 311 times 23 times 100. This tells us to multiply 311 by 23, which is our previous example, and which gives us the already known answer of 7,153. Then we still have to multiply by the 100. But multiplying any number by 100 merely has the effect of placing two zeroes at the end of the number. So we put two zeroes at the end of 7,153, and the result is 715,300, as before.

The advantage of this fourth method is that it tells us what to do in certain other cases also. Suppose that the two zeroes were at the end of the 23. We are multiplying 311 by 2,300. The same reasoning as in Paragraph 4, tells us that these two zeroes must go at the end of the answer. Again we have 715,300. In fact, if one zero were on the 311 and the other on the 23, so that we multiply 3,110 by 230, both zeroes go at the end of the answer, and again we have 715,300.

RULE: Collect all zeroes at the end of the multiplicand and at the end of the multiplier and put them at the end of the answer. Then go ahead and multiply without paying any further attention to them.

For instance, this was our first example:

$$\underline{0\ 0\ 2\ 3} \quad \times \quad 1\ 4$$
$$3\ 2\ 2$$

Suppose we wished to multiply 230,000 by 140; what would the answer be? Simply do the example as before, without the zeroes, then collect all the five terminal zeroes and write them after your answer:

$$\underline{0\ 0\ 2\ 3} \quad \times \quad 1\ 4$$
$$3\ 2\ 2\ 0\ 0\ 0\ 0\ 0$$

The answer is 32,200,000.

THREE-DIGIT MULTIPLIERS

So far, we have been multiplying various numbers by multipliers consisting of only two digits. The multiplicand may have been long, like the 241,304 of one example, but the number it was multiplied by was a number of two figures; in the example it was 32. How shall we multiply various numbers by a three-digit multiplier?

Let us take an illustration: 213 times 121. The multiplier has three digits, so we write three zeroes in front of the multiplicand:

$$0\ 0\ 0\ 2\ 1\ 3 \quad \times \quad 1\ 2\ 1$$

This agrees with the rule that we mentioned before, of placing in front of the left-hand number, or multiplicand, as many zeroes as there are digits in the multiplier (sometimes one of these zeroes will be wasted, as happened before). Then we take the work in steps, finding one figure of the answer at each step.

First step: $0\ 0\ 0\ 2\ 1\ \underline{3} \quad \times \quad 1\ 2\ \underline{1}$

answer: 3 — *3 times 1 is 3*

Second step: $0\ 0\ 0\ 2\ \underline{1}\ \underline{\underline{3}} \quad \times \quad 1\ \underline{\underline{2}}\ \underline{1}$

answer: 7 3

work: 1×1
$+$
3×2

(figures with a single underline are multiplied together; likewise those with a double underline).

So far, in these first two steps, we have done only what we did in the preceding section. The calculation so far is the same as if we were multiplying 13 by 21, instead of 213 by 121.

Third step: This is new: we obtain the next figure of the answer by adding together *three* pairs of numbers, instead of two:

$$0\ 0\ 0\ \underline{2}\ \underline{\underline{1}}\ \underline{\underline{\underline{3}}} \quad \times \quad \underline{\underline{\underline{1}}}\ \underline{\underline{2}}\ \underline{1}$$

answer: 7 7 3

work: 2×1
$+$
1×2
$+$
3×1

The way the "work" is done can be seen from the manner in which the numbers have been printed. To make it clearer we can draw curves for the "outside" and the "inside" pairs as we did before, but now we have a "middle" pair as well:

0 0 0 2 1 3 × 1 2 1

The outermost line connects the 2 of 213 with the last 1 of 121. So the outside pair gives 2 times 1 as its contribution to the answer. We move inward. The middle line connects the 1 of the 213 with the 2 of 121, and we have 1 times 2. That is the second contribution to the answer. The third and last contribution comes from the innermost line which gives us 3 times 1. Adding these three contributions, we have 2 plus 2 plus 3, which is 7, the next figure of the answer. The outermost pair, 2 and 1, is identified by the same rule as before: the figure of the multiplicand directly above the next space where the answer is to be found is part of the outside pair. The other figure of this pair is the last digit of 121. The figures next to these form the middle pair and the remaining figures are, of course, the inner pair.

The rest of the work consists of repeating this step with three curved lines, but moving the lines over toward the left:

Fourth step: 0 0 0 2 1 3 × 1 2 1
5 7 7 3

work: 0×1
\+
2×2
\+
1×1

the 5 of the answer is zero plus 4 plus 1

Last step: 0 0 0 2 1 3 × 1 2 1

2 5 7 7 3

work: 0×1
$+$
0×2
$+$
2×1

the 2 of the answer is zero plus zero plus 2

This is the last step in this particular example because we have just found the figure 2 with nothing to carry. There is a zero in front of the answer for the sole purpose of taking care of a carried figure at the last step, but now we have nothing to carry so we have finished. The answer is 25,773.

Of course, this example has seemed rather long simply because it was stretched out with detailed explanations. In actual practice it goes fast. Following is one in a form closer to actual work. Underlining the figures takes the place of pointing with the fingers and "moving inward" to mark the outside and inside pairs. The two figures with a single line under them are to be multiplied together, those doubly underlined are to be multiplied, and those triply underlined are to be multiplied:

First step: $0\ 0\ 0\ 3\ 0\ \underline{2} \quad \times \quad 1\ 1\ \underline{4}$

8

Second step: (Of course, in a practical problem we would not write the numbers over again!)

$0\ 0\ 0\ 3\ \underline{0}\ \underline{\underline{2}} \quad \times \quad 1\ \underline{\underline{1}}\ \underline{4}$

2 8 *zero plus 2 is 2*

Third step: $0\ 0\ 0\ \underline{3}\ \underline{\underline{0}}\ \underline{\underline{\underline{2}}} \quad \times \quad \underline{\underline{\underline{1}}}\ \underline{\underline{1}}\ \underline{4}$

·4 2 8 *12 plus zero plus 2 is 14*

Fourth step: $0\ 0\ \underline{0}\ \underline{\underline{3}}\ \underline{\underline{\underline{0}}}\ 2 \quad \times \quad \underline{\underline{\underline{1}}}\ \underline{\underline{1}}\ \underline{4}$

$4\ {}^{\bullet}4\ 2\ 8$ — *zero plus 3 plus zero plus the dot is 4*

Last step: $0\ \underline{0}\ \underline{\underline{0}}\ \underline{\underline{\underline{3}}}\ 0\ 2 \quad \times \quad \underline{\underline{\underline{1}}}\ \underline{\underline{1}}\ \underline{4}$

$3\ 4\ {}^{\bullet}4\ 2\ 8$ — *zero plus zero plus 3 is 3*

This is the last step because there is nothing carried (no dot). If there were a dot we would put a 1 under the left-hand zero. If there were a double dot (a carried 2), we would put a 2 under the leftmost zero. As it is, however, the calculation is finished and the answer is 34,428.

Remember, zero times any number is zero.

Finally, you can see how easily a problem goes in actual work by doing one yourself:

$$0\ 0\ 0\ 2\ 0\ 3 \quad \times \quad 2\ 2\ 1$$

You start by multiplying the 3 of 203 by the 1 of 221, as you no doubt did. Then you multiply the 0 3 against the 2 1 of 221 in inside and outside pairs and add the two results; and so on. The answer, as you have probably discovered, is 44,863.

For practice, if you wish, you can make up examples of your own and work them. It will make it easier if you do a few with two-digit multipliers first, like 23 or 31, and then afterward use three-digit multipliers.

This method can be used successfully with numbers of any length. However, when a long number contains many large digits, as 9,869 does, we shall find ourselves "carrying" rather large numbers when using the method of this chapter. That is why we favored the smaller digits like 2 and 3 in the examples we have been doing. If you wish to try one or two problems with large digits you will be in a better position to appreciate the "speed multiplication" of the next chapter, which will require us to carry only 1's and 2's. In the meantime, the method of this chapter works very conveniently on small-digit numbers and, in addition, *it is an indispensable part of the next chapter.*

MULTIPLIERS OF ANY LENGTH

For longer multipliers we use the same principles. Four-digit multipliers, like 3,214, work in this way: each figure of the answer is found by adding up four pieces. Each of these four pieces is the result of multiplying two digits together. What two digits? Those at the ends of a curved line; in other words, outside and inside pairs. Take the example 2,103 times 3,214:

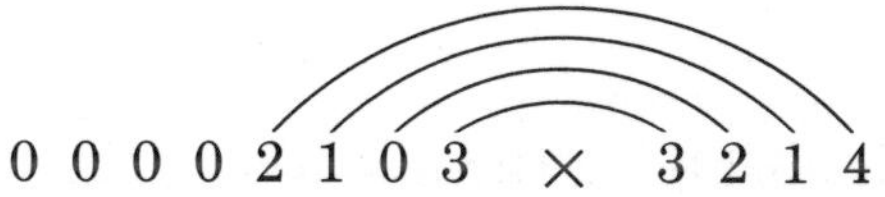

The picture shows the four pairs that we would use at a stage halfway through the calculation. At that stage we would say "2 times 4 is 8, plus 1 times 1 is 9, plus zero is 9, plus 9 is 18." There are *four* zeroes in front because 3,214 consists of *four* figures. The leftmost zero will not be needed unless there is a carried figure at the last step; it is there just in case. Here is how the work would go:

First step: $0\ 0\ 0\ 0\ 2\ 1\ 0\ \underline{3} \times 3\ 2\ 1\ \underline{4}$

$\cdot 2$ — *3 times 4 is 12*

Second step: $0\ 0\ 0\ 0\ 2\ 1\ \underline{0}\ \underline{\underline{3}} \times 3\ 2\ \underline{\underline{1}}\ \underline{4}$

$4\ \cdot 2$ — *zero plus 3 plus the dot*

Third step:	0 0 0 0 2 1 0 3	×	3 2 1 4
	·0 4 ·2		*4 plus zero plus 6*
Fourth step:	0 0 0 0 2 1 0 3	×	3 2 1 4
	·9 ·0 4 ·2		*8 plus 1 plus zero plus 9 plus the dot*
Fifth step:	0 0 0 0 2 1 0 3	×	3 2 1 4
	5 ·9 ·0 4 ·2		*zero plus 2 plus 2 plus zero plus the dot*
Sixth step:	0 0 0 0 2 1 0 3	×	3 2 1 4
	7 5 ·9 ·0 4 ·2		*zero plus zero plus 4 plus 3*
Seventh step:	0 0 0 0 2 1 0 3	×	3 2 1 4
	6 7 5 ·9 ·0 4 ·2		*zero plus zero plus zero plus 6*

There is nothing to carry, and the pairs of digits will all give zero, so we have finished. The answer is 6,759,042. From this, it is clear how we would proceed with multipliers of any length.

SUMMARY

In this chapter we have worked with two-digit numbers multiplied by two-digit numbers, like 31 times 23; then with larger multiplicands and two-digit multipliers, like 32,405 times 42; and then with both multiplicands and multipliers of any length, like 32,405 times 422. In all these cases, we get the right-hand figure of the answer by multiplying together the right-hand figures of the two numbers. In all cases, we get the middle figures of the answer by using outside and inside pairs and adding the results. Finally, we can get the left-hand figure or figures of the answer by writing zeroes in front of the multiplicand and applying the outside and inside pairs to these zeroes; we write as many zeroes as there are digits in the multiplier.

You can test yourself if you wish, and at the same time make the procedure more vivid in your mind, by doing the following practice examples:

1. 31 × 23
2. 33 × 41
3. 63 × 52
4. 413 × 24
5. 224 × 32
6. 705 × 25
7. 511 × 61
8. 341 × 63
9. 4133 × 212
10. 31522 × 3131

Answers: **1.** 713 **2.** 1,353 **3.** 3,276 **4.** 9,912 **5.** 7,168 **6.** 17,625 **7.** 31,171 **8.** 21,483 **9.** 876,196 **10.** 98,695,382

CHECKING THE ANSWER

The following method of checking our answer was not invented by Professor Trachtenberg, but was incorporated into his system because it is so simple and convenient. It has been known to mathematicians for hundreds of years, yet it is not widely known to the layman and seems to be little used in everyday life. For that reason we shall explain it under the name of the "digit-sum method." The essence is that we add the digits across each number, as in the next paragraph.

A digit is any one of the single-figure numbers, 1 through 9. Zero is also a digit. So any number at all is made up of certain digits. The "digit-sum" is what we get when we add the digits across the number, like this:

Number: 4 1 3
Digit-sum: 4+1+3=8

But hereafter, we shall always understand that the digit-sum has been reduced to a single figure by adding across again when necessary. For instance, suppose the number is 6,324. We shall have:

Number: 6 3 2 4
Digit-sum: 6+3+2+4=1 5
This 15 gives: 1+5=6

So the digit-sum of 6,324 is 6. In other words we shall work with reduced digit-sums. We do so because it will make the work easier in the next step.

As a check in multiplication we need to find three digit-sums: the digit-sum of the multiplicand, the digit-sum of the multiplier, and the digit-sum of the product. For instance, suppose we have done this multiplication and wish to check it:

$$\frac{0\ 0\ 2\ 0\ 4}{6\ 3\ 2\ 4} \times 3\ 1$$

Three numbers are involved, the two that are multiplied and the answer. We find the digit-sum of each one:

	Number	*Digit-sum*	
Multiplicand:	2 0 4	6	
Multiplier:	3 1	4	
Product:	6 3 2 4	6	*because 15 reduces to 1 plus 5 is 6*

The rule for checking the work is this:

The digit-sum of the product should be equal to the digit-sum of the product of the digit-sums of the multiplier and multiplicand.

If they are not equal there is something wrong. In our example the digit-sum of the product, 6,324, is 6. This ought to be equal to the digit-sum of the product of the other two digit-sums. Is it? We can tell by multiplying: 6 times 4 is 24, which reduces to 6. We have 6 again, so the work checks.

The multiplication of these digit-sums is always very easy, because they are only single figures. The check goes along parallel to the original multiplication:

numbers:	2 0 4	×	3 1	=	6 3 2 4
digit-sums:	6	×	4	=	24 (that is, $2+4$) $= 6$

Short-cuts

We can save ourselves some trouble in adding the digits across a number to find the digit-sum. It is especially worthwhile in finding the digit-sum of a very large number. Here are the timesavers:

(1) Reduce to a single figure as you go along, don't wait until the end. Suppose you are finding the digit-sum of 252,311. Start at the left and add across: 2 plus 5 plus 2 and so on. Say to yourself only the totals, which are running totals: 2, 7, 9, 12 . . . and now "reduce to a single figure as you go along." Reduce this 12 to 3 (1 plus 2). Go ahead with this 3 and add to it the remaining two digits of our example: 3, 4, 5. The digit-sum is 5. This is less trouble than adding 2 plus 5 plus 2 plus 3 plus 1 plus 1 equals 14, then 1 plus 4 equals 5. In a very long number it can save a good deal of time. For instance, the digit sum of 6,889,567 is 4. Reducing as you go along, you would say to yourself "6, 14, *is 5,* 13, *is 4,* 13, *is 4,* 9, 15, *is 6,* 13, *is 4.*" Otherwise, you would have to add all the way up to the 49.

(2) Disregard 9's. If the number that you are adding across happens to contain a 9, or several 9's, pay no atten-

tion to them, leave them out of the addition. You will reach the same result as if you had added them in. This may seem odd, but it is always true. The digit-sum of 9,399 is 3; we ignore the 9's. If you add them in you will have a total of 30, then reduce it: 3 plus zero is 3. Furthermore, if you happen to notice two digits that add up to 9, you may ignore both of them: the digit-sum of 81,994 is 4, because 8 plus 1 is 9, and the 9's don't count. It is only safe to do this if the two that add to 9 are touching, or at least close together. If they are not, you may forget that you have decided to ignore them and add one of them in.

This method of checking will be useful in the next chapter. It can also be used on any practice problems that you may make up for yourself on the method of this chapter, and it will be used in calculations other than multiplication.

CHAPTER THREE

Speed multiplication— "two-finger" method

As we saw in the last chapter, an important advantage of the Trachtenberg system is that we are able to multiply any number by any other number and write down the answer immediately. We do not write down the intermediate figures as in conventional multiplication. The direct method we have just learned is one of general application—it can be used in multiplying any two numbers together. But, in many cases, it needs a further improvement which is the subject of this chapter. When we have numbers made up mostly of the larger digits, as in 978 times 647, we are likely to have large numbers to add up mentally and large numbers to carry. The further improvement of the method consists of eliminating these inconveniently large numbers from our mental work. We do this by adding to the method a new feature—what Professor Trachtenberg called the "two-finger" method. It might equally well be called the "units-and-tens" method. You will see where the names come from as we go into the method, because both are descriptive.

We are going to look at the new feature by itself, first of

all, then we shall apply it to the problem of doing a full-sized and practical multiplication. So for the moment, we put out of our minds the process of multiplication that we have been working on and we concentrate on the following points:

1. A digit is a one-figure number, like 5 or 7. Zero is a digit.

2. When we multiply a digit by a digit, we get a one-figure or a two-figure number, never any longer. Proof: the largest number we can have by multiplying digits is 9 times 9, or 81, which is only a two-digit number.

3. Sometimes a digit times a digit gives a one-figure answer, like 2 times 3. In those cases the resulting 6, or whatever it is, will be treated in this system as a two-figure number by writing a zero in front of it. We shall say, in the method we are about to exhibit, that 2 times 3 is 06. This has the advantage of simplifying the rules and procedures, by standardizing all products of digits as two-figure numbers. Of course, writing a zero before a number, as in 06, has no effect on the actual value of the number.

4. In any two-figure number, the left-hand figure is the "tens" figure and the right-hand figure is the "ones" or "units" figure. For instance, in the number 37, the tens-figure is 3 and the units-figure is 7. This agrees with our everyday usage, because if we have 37 dollars, we have the equivalent of 3 tens and 7 ones, not the other way around.

5. In using our new method, we shall often have occasion to use only the units-digit of a number. For instance, we may come across the number 24 and say merely "4," ignoring the 2 that is the tens-digit. This sounds as though it might cause errors, dropping out a figure and forgetting it. However, it works out because the forgotten tens-digit comes into its own somewhere else. There are other occasions where we use only the tens-digit and forget the units-digit. In such case we would look at 24 and say "2."

6. THIS IS IMPORTANT. In the new method we very frequently have to combine the idea of point **2** with idea of point **5**. That is to say, in multiplying two digits together (like 3 times 8) we would use only the units-digit of the result (4 of the 24 that we get from 3 times 8) or, in just as many other cases, only the tens-digit. For instance, we might have 5 times 7 and use only the 3, the tens-digit of 35.

This is an unfamiliar mental operation. We do not have it in conventional multiplication, and it has a rather strange feel the first few times that we try it. See for yourself by looking at these examples and saying to yourself only the units-digit of the products:

(1) 4×3 (2) 3×6 (3) 5×4 (4) 8×2

Answers: 2, 8, 0, and 6. Now go back over the same examples and say to yourself only the tens-digit in each case. The answers are, of course, 1, 1, 2, and 1.

7. Here is where the "units-and-tens" name really comes in. Put two digits side-by-side, like 3 and 8. Multiply each of them by another digit, say 4, using the idea of taking only the units- or the tens-digit of the result, as we did in point **5**. We do this in a particular way, however. We use the units-digit only when we multiply the left-hand figure (the 3) and the tens-digit for the right-hand figure (the 8). The U means that we keep only the units of the result, the T means that we keep only the tens. The result, with the dropped-out figures in parentheses, is:

$$\begin{array}{cccc} \text{U} & \text{T} & & \\ 3 & 8 & \times & 4 \\ \hline (1)2 & 3(2) & & \end{array}$$

We shall always have the U and T in that order, from now on. With the left-hand figure of the adjacent pair, like the

3 of 3 8, we use only the units-digit of the product. On the right-hand figure of the pair, the 8 of 3 8, we use only the *tens*-digit of that product.

8. Finally, we make one more very simple step—we *add* the two figures that we found in point **7**. We found 2 and 3, in the example. Now add them, and get 5. This is the product we shall use in doing actual multiplication.

Notice that we obtained only the single figure 5, out of the pair of digits 3 8. We "multiplied" the 3 8 by 4, but it was not ordinary multiplication. This is a characteristic of the "units-and-tens" method: a pair of digits is multiplied by a third digit and you end up with only one digit, like the 5 of the example. It happens because we use only the units of one result, throwing away the 10's, and vice versa with the other.

Because this is the essential part of the "two-finger method," we will show this example in full. We use again 3 8 times 4 (not "times" in the ordinary sense!), and we exhibit the relations in the following diagram:

$$\begin{array}{rccl} & \text{U} & \text{T} & \\ \text{problem:} & \underline{3} & \underline{8} & \times \quad 4 \\ \text{work:} & 1\underline{2} & \underline{3}2 & \\ & \underline{2} & \underline{+3} & \\ \text{answer:} & \multicolumn{2}{c}{5} & \end{array}$$

We hasten to add that nothing so complete is needed in actual practice, after the method has become familiar. Nothing is written except the numbers to be worked on, that is, 3 8 and 4, and the result, 5. More than that: after the method is learned you should try to avoid even *thinking* the explanatory numbers that you see in the diagram. It should be a semi-automatic mental process, most of it going on below the fully conscious level. You look at the 3 8 and the

4 and are half aware of the 2 and 3 (in 12 and 32), and you say "5" to yourself almost immediately. That high a degree of facility will come after a good deal of practice, just as in any other skill.

To emphasize the importance of this process we call the result by a special term, the "pair-product." The 5 that we got in the above example is the "pair-product" of 3 8 × 4.

> **Definition.** A pair-product is a number obtained by multiplying a pair of digits by a separate (multiplier) digit in this special way: we use the multiplier-digit to multiply each digit of the pair separately, and then we add together the units-digit of the product of the left-hand digit of the pair and the tens-digit of the product of the right-hand digit of the pair.

The figure that we get in this way, the pair-product, is of use to us because it enables us to perform rapid multiplication without carrying large numbers, or even encountering large numbers. How this comes about we shall see in a moment. First let us look at some examples which will bring out a couple of minor points of interest:

U	T		
8	2	×	5

The answer is 1, as you have probably noticed; if not, you can see it from this:

	U	T			
problem:	8	2	×	5	
work:	40 (0 underlined)	10 (1 underlined)			*8 times 5 is 40; 2 times 5 is 10*
	0	+1			
pair-product:	1				

```
               U  T
     problem:  4  1   ×  3
        work: 12  03        4 times 3 is 12; 1 times 3
               2+0          is 3 or 03
pair-product:   2
```

```
               U  T
     problem:  1  4   ×  3
        work: 03  12        1 times 3 is 03; 4 times 3 is 12
               3+1
pair-product:   4
```

```
               U  T
     problem:  2  8   ×  4
        work: 08  32        2 times 4 is 08; 8 times 4 is 32
               8+3
pair-product:  11 or ·1
```

We add two digits together to get the pair-product. This can cause us to go over 10, as we had 11 in the last example. But notice that this will not take us over 18, so a single dot will take care of it. The points illustrated in the examples were these:

1. Try to think of a single-digit product as if it had a zero written in front. For instance, 2 times 2 is 04, and 6 times 1 is 06. The purpose of this is to guard against a human tendency to error. Thinking quickly of the tens-digit of 2 times 2, we may be misled by the 4 that pops into our minds.

2. When we add the two partial products, the tens of one and the units of the other, the two together may sometimes bring us over 10; that is, it may bring us into two-digit numbers. In that case we do as usual, we rewrite the units-digit (the 3 of 13, for instance) and we indicate the tens-digit (the 1 of 13) by a dot. This means that we are doing some carrying. But it is an easy kind of carrying. We shall not need to carry 15, as we would in some other kinds of multiplication when we come to a total of 153 for one figure of the answer. The smallness of the carried figure is important *because it indicates the smallness of the figures that you were working with.*

3. Remember always, zero times any number is zero, but multiplying any number by *one* leaves the number unchanged.

4. Only one or two of the pair-product calculations should be done in full written form, as we diagrammed 3 8 by 4 a few pages back. After one or two written ones, you should make the necessary effort of concentration to visualize the two numbers—like $1\underline{2}$ and $\underline{3}2$—and combine the inner digits (to get 5). It is easy enough to visualize this mentally, even before we have practiced it. The important thing is to get this down pat. We wish to get to the point where we feel that we are leaving out some of the steps—which means really that we are doing some of the steps of the calculation without focusing our attention on them. Practice will bring us to this point.

Try a few more problems, with the points just mentioned in mind:

U T
6 4 × 3

U T
3 5 × 7

U T
6 3 × 5

U T
7 5 × 7

U T
2 6 × 3

U T
7 2 × 5

U T
9 4 × 3

U T
4 1 × 8

```
U T                                   U T
6 6  ×  5                             1 6  ×  6
```

Stop here a moment. Is everything perfectly clear up to now? If not, it will be worth while to go back and re-read whatever is needed. The ideas that we have been considering in the last few pages are not complicated, but it is important to have them absolutely clear. They are the heart of the method of this chapter.

MULTIPLICATION BY A SINGLE DIGIT

We can use the pair-products that we have been studying to perform simple multiplications. Suppose we wish to multiply 3,112 by 6. This, of course, is an easy example, but we will start with easy ones and work up to the more difficult ones. Using the pair-products, we have a new way to perform this multiplication. The basic idea is this:

Each pair-product is one figure of the desired answer.

Let us do an example in full. We set it up with a zero in front, as we did in the last chapter. We put the U of the UT over the position where the "next" figure of the answer —now the first figure—will appear:

```
        U T
0 3 1 1 2  ×  6
```

The T has no work to do now, because it is not over a digit. We simply use the units of 2 times 6.

First step:

```
        U T
0 3 1 1 2  ×  6
---------
        2        this 2 is the units-digit of 12
```

Second step:

$$\begin{array}{ccccccc} & & & \text{U} & \text{T} & & \\ 0 & 3 & 1 & 1 & 2 & \times & 6 \\ \hline & & & 7 & 2 & & \end{array}$$

The UT has moved to the left. That is because the U of the UT is always placed over the position where the next figure of the answer is to appear, that is, where the 7 will be. In this example, the 7 is the pair-product of the units-digit of $0\underline{6}$ (from 1 times 6) plus the tens-digit of $\underline{1}2$ (from 2 times 6).

In the first step, the 2 of 3112 was used as a U. In the second step it was used again, but this time as a T. This always happens. Each figure of the multiplicand is used twice, once under the U of UT and then under the T.

Third step: Move the UT to the next figure of the multiplicand.

$$\begin{array}{ccccccc} & & \text{U} & \text{T} & & & \\ 0 & 3 & 1 & 1 & 2 & \times & 6 \\ \hline & & 6 & 7 & 2 & & \end{array}$$

This 6 is the pair-product of the units-digit of $0\underline{6}$ (1 times 6 is 06), plus the tens-digit of $\underline{0}6$ (again 1 times 6 is 6).

Fourth step: Move to the next figure of the multiplicand.

$$\begin{array}{ccccccc} & \text{U} & \text{T} & & & & \\ 0 & 3 & 1 & 1 & 2 & \times & 6 \\ \hline & 8 & 6 & 7 & 2 & & \end{array}$$

The 8 is the pair-product of the units-digit of $1\underline{8}$ (3 times 6), plus the tens-digit of $\underline{0}6$ (1 times 6 is 06).

Fifth step: Move to the last figure, the zero in front of the multiplicand.

$$\begin{array}{ccccccc} \text{U} & \text{T} & & & & & \\ 0 & 3 & 1 & 1 & 2 & \times & 6 \\ \hline 1 & 8 & 6 & 7 & 2 & & \end{array}$$

This is the pair-product of the tens-digit of $\underline{1}8$ (3 times 6), plus the units digit of $0\underline{0}$ (zero times 6 is zero).

Obviously, when we come to a zero in the long number, we don't need to think about units-digit or tens-digit. The act of multiplying by zero annihilates the 6.

In this method there is no reason to avoid large digits as was advisable in the direct method. Let us look at a problem with large digits, instead of merely the 1, 2, and 3, of the easy examples. We are familiar now with the UT above the pair of digits, and we may omit them. On the other hand there is still some danger of losing our place and picking up the wrong digit. As a compromise, we now show a curved line that is forked at one end, enabling it to point to both digits of the digit-pair:

First step: 0 7 5 8 – × 7

6 — the U of 56— the dash constitutes nothing, no tens

Second step: 0 7 5 8 – × 7

·0 6 — the ·0 (10) is the U of 35 (5 times 7) plus the T of 56 (8 times 7)

Third step: 0 7 5 8 – × 7

·3 ·0 6 — the ·3 or 13, is the U of 49 (7 times 7) plus the T of 35 (5 times 7) plus the dot from ·0

Fourth step:	0 7 5 8 – ×	7
	5 ˙3 ˙0 6	the 5 is the tens-digit of 49 (7 times 7), plus the dot from ˙3; the zero from 0 times 7 contributes nothing

So the answer is 5,306.

In this example we did not receive the full benefit of the power of the units-and-tens method. We could have done the problem easily, even with conventional multiplication. The point is that this happens only with the simplest examples. In most cases where we find ourselves needing to do a multiplication, the numbers we have to deal with will not be specially selected to make the work easy. The units-and-tens method takes care of all kinds of problems.

We shall see the advantage of the units-and-tens method when we come to multiplication by larger multipliers, not merely by single digits like 6 or 7. In the meantime, this multiplication by single digits is excellent practice because a long multiplier is made up of single digits, and the calculation is an extension of what we are doing now.

Incidentally, if you look back at the problem we have just done and observe how the forked line moves to the left across the multiplicand, you can understand how the name "two-finger" method originated. A person who is first beginning to familiarize himself with the method may have some trouble in keeping his place; that is, in remembering which pair of digits he is multiplying and which one has the units designation. To keep track of the process as he goes along, he may point at the two figures of the digit-pair with the forefinger and the middle finger of the left hand. Call the middle finger of the left hand the "units finger," and the forefinger the "tens finger." Then, once he has identified his fingers, he can always keep his place in a calculation by pointing at the digit-pair. The middle finger takes the place of the letter U that we have written above the numbers, and the forefinger takes the place of the T. If you find it helps you, by all means try it. In any case, you will find quite soon that you can dispense with the pointing because you know where you are.

On the other hand, even after you find that you can get along without the pointing fingers or the curved lines, you should still make it a habit to set out the work in a neat and orderly fashion. The fact that the work is well arranged is a simple safeguard against the human tendency to make careless errors, especially when one is working at high speed. The point is so true, and so obviously true, that one might think it would hardly be necessary to mention it. Unfortunately, experience shows that most people, even the majority of quite intelligent persons, will write the work out in a crowded and irregular fashion.

Try your hand at one or more of these examples, keeping your place in any way you find convenient:

1. 5 6 × 8
2. 5 6 7 × 9
3. 8 5 4 × 4
4. 8 4 5 6 3 × 6

Answers: **1.** 448 **2.** 5,103 **3.** 3,416 **4.** 507,378

Make up examples of your own and work them. The more you do of these simple ones, the faster and easier you will do the long and difficult multiplications. These single-digit problems are the foundation of the fast multiplication method.

MULTIPLICATION BY TWO-DIGIT NUMBERS

We have been multiplying numbers of any length by single figures, like 6 or 7. But how shall we multiply a long number by 37, for instance? Or by 2,237? We begin by extending our method to the two-digit multipliers first.

The over-all idea is that we now add the method of the previous chapter to what we have been doing in this chapter. This is true not only for two-digit multipliers, but for all multipliers. If you will refresh your memory on how we worked the outside and inside pairs, and moved across the numbers, you will see the same thing happening here. The difference is that now we are also using the units-and-digits feature.

Consider the problem 73 times 54. In the last chapter we worked problems like this by the outside and inside pairs. Let's not bother to do the multiplication by the method of the last chapter but, for the sake of comparison, we shall indicate the way that the figures pair themselves off. X marks the place of the figure of the answer that we get at each step:

First step: 0 0 7 3 × 5 4
(X under the 3)

DON'T WORK IT! just notice the position of the curves

Second step: 0 0 7 3 × 5 4
(X under the 7)

Third step: 0 0 7 3 × 5 4
(X under the second 0)

Fourth step: 0 0 7 3 × 5 4

X *this X is the last carry*

That was how we did it in the last chapter. Now compare it with the diagrams for the improved method. Here again we are not going to work the problem, but merely observe the movement of the pair lines:

First step: 0 0 7 3 – × 5 4

X *The broken lines will be needed later, but not in this step.*

Have you noticed that the figure of the multiplicand directly above the space where the next figure is to appear is still part of the outside pair?

Second step: 0 0 7 3 – × 5 4

Here is where the units-and-tens feature comes into play. Each figure of the multiplier works on *two* figures of the multiplicand. In this example, the 4 of the multiplier acts on the adjacent digits 7 3 in the multiplicand to obtain a pair-product. The solid connecting line indicates the units-figure; the broken connecting line, the tens-figure. That is, we use the units of 7 times 4 and the tens of 3 times 4:

U T

0 0 7 3 × 5 4

2<u>8</u> + <u>1</u>2

Before we finish the calculation, see how the lines move across the multiplicand exactly the same as they moved in the method of the preceding chapter:

Third step:

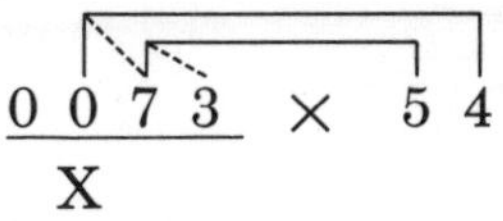

The 4 of 54 is now connected with both zero and 7; in addition, the 5 is connected with 7 and 3.

Fourth step:

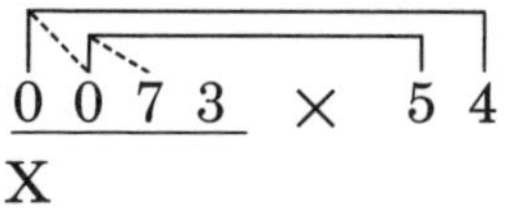

IMPORTANT: The position of the lines is of the utmost importance in this method. It is the whole secret of getting the right answer. The rest of it, multiplying the digits together and taking the units or tens of their product, is easy. Besides, you have already mastered it. In doing difficult multiplications with long numbers, the difficulty comes entirely in using the *correct* two digits to match as a pair and multiply together. The right pair of digits, at the right time, will give you the next figure of the answer.

As in the method of the preceding chapter, the figure of the multiplicand directly above the space where the answer is to appear is part of the outside pair; to be specific it is now the units-figure of the outside pair; the tens-figure is the digit to its immediate right. From here we work inwards, locating the units- and tens-figures of the inside pair.

To find the next figure of the answer, you may either draw or imagine drawn the leftmost line of the little pattern of lines that we saw in the diagram above. This will be directly above the next figure of the answer to be calculated. Then you can easily visualize the rest of that pattern, if you have looked at the diagram and understood how it goes. The example in full is as follows:

U T

First step: 0 0 7 3 - × 5 4

2 — *the units of 3 times 4 is 2; other lines touch nothing*

U T
U T

Second step: 0 0 7 3 - × 5 4

·4 2 — *the 4 of 54 gives $2\underline{8}$ plus $\underline{1}2$ is 9; the 5 gives $1\underline{5}$ plus nothing is 5; total is 14*

Perhaps if we simplify the picture the main point will show up more clearly, like this:

	U T ←
	U T ←
	0 0 7 3 - × 5 4
answer	·4 2
4 gives	$2\underline{8}+\underline{1}2$
5 gives	$1\underline{5}+0$

adding the underlined units- and tens-figures, 8 plus 1 plus 5 plus 0 is 14

The meaning of the diagram is that we make the 4 of 5 4 act on the digit pair 7 3, like this:

U T
7 3 × 4

Then we make the 5 of 5 4 act on the pair 3 and nothing:

U T
3 - × 5

The 7 3 × 4 gives us 9 (because $2\underline{8}$ plus $\underline{1}2$ is 9) and the 3 - × 5 gives us 5 ($1\underline{5}$ plus nothing), in the usual units-and-tens way that we have worked before. Adding these two results, 9 plus 5 is 14, we write the 4 and put a dot for the 1 of 14.

All this should be done mentally in actual work. We write out everything here only for the sake of explanation. It will be easy to do these steps mentally after we have practiced a while on the units-and-tens method. That is why we said, a few pages back, that practice in multiplying by single digits is invaluable preparation for the whole method.

```
              U  T←———————————┐
                 U  T←———┐    │
Third step:   0 0  7  3   ×   5 4
   answer       9 ·4  2
   4 gives      0+28
   5 gives         35+15
```

$0+2\underline{8}$; $3\underline{5}+\underline{1}5$

adding the indicated units and tens, we get the result 0 plus 2 plus 5 plus 1 is 8, and the dot makes the answer 9

```
              U T
                U T
Fourth step:  0 0 7 3   ×   5 4
              3 9·4 2
   4 gives    0 0
   3 gives      0 35
```

$0\ \underline{3}5$

when we multiply by zero we don't need to write the result as 0 0, obviously we are not going to get anything: 0 plus 0 plus 0 plus 3 is 3

So the answer is 3,942.

LONG NUMBER BY TWO-DIGIT MULTIPLIER

The same process enables us to multiply a number of any length by two digits. We had, just now, 73 times 54. Suppose we wish to multiply 5,273 by 54. It begins with exactly the same first two steps:

```
                  U T ←──┐
First step:   0 0 5 2 7 3  ×  5 4
              ───────────
                      2        4 gives 12 plus 0
```

```
                U T ←────────┐
                  U T ←──┐   │
Second step:  0 0 5 2 7 3  ×  5 4
              ───────────
                    ·4 2       4 gives 28 plus 12; 5 gives
                               15 plus 0; underlined figures
                               add to 14
```

(In the notes above, the underlined figures are the 2 of 12, the 8 of 28, the 1 of 12 and the 5 of 15.)

Now we continue with the same process:

```
                 U    T ←─────────────┐
                      U    T ←────┐   │
Third step:   0 0 5 2    7    3  ×  5 4
              ─────────────────
                   ·7   ·4    2
              ─────────────────
   4 gives         08+28
   5 gives              35+15        underlined figures add
                                     to 16, plus the dot is 17
```

```
               U    T ←─────────────┐
                    U    T ←────┐   │
Fourth step:  0 0 5    2    7 3  ×  5 4
              ─────────────────
                  4   ·7   ·4 2
              ─────────────────
   4 gives      20+08
   5 gives           10+35           underlined figures add
                                     to 3, and the dot of 17
                                     makes it 4
```

(Underlined: the 8 of 08 and the 2 of 28, the 5 of 35 and the 1 of 15 in the third step; the 0 of 20 and the 0 of 08, the 0 of 10 and the 3 of 35 in the fourth step.)

```
              U T ←――――――――――┐
                U T ←―――――┐  |
Fifth step:   0 0 5 2 7 3  ×  5 4
                8 4·7 4 2                work this by yourself
```

```
              U   T ←――――――――――――――┐
                  U   T ←――――――――┐  |
Last step:    0   0   5 2 7 3  ×  5 4
              2   8   4·7 4 2
              0+0
                  0+25
```

The answer is 284,742.

In actual work we would certainly not re-write the figures all over at each step—once is enough!—and we would not write any of the "work" figures. They can easily be done mentally. The calculation goes fast and is easy, if we have practiced with pair-products. Seeing 7 3 times the 4, of say 5 4, we should say "9" almost immediately. The picture 28 +12 will require no mental effort after we do a reasonable number of them as practice. In fact, as we mentioned before, it will become a semi-automatic mental operation, and will be done without focusing full attention. This is the way it should be.

THREE-DIGIT MULTIPLIERS

In doing the multiplication of a number of any length by a multiplier consisting of three digits, such as 273 times 154, or 5,273 times 154, or 235,273 times 154, the same general principles apply. Now, however, *each figure of the answer is the sum of three parts.* Before, it was the sum of two parts. Each of these three parts comes from a different digit-pair, as we shall see, by using the usual units-and-tens idea. Let us look at an example, 273 times 154. We put 3 zeroes in front (154 has 3 digits!).

First step:

U T ← (4)

0 0 0 2 7 3 × 1 5 4

2

4 of 154 gives $1\underline{2}+0$

the 5 and the 1 of 154 do not pair with anything, so they do not contribute to the answer

Second step:

U T ← (4)

U T ← (5)

U T ← (1)

0 0 0 2 7 3 × 1 5 4

•4 2

4 gives $2\underline{8}+\underline{1}2=9$

5 gives $1\underline{5}+\underline{0}=5$

the 1 gives nothing, because it fails to pair with anything

In fact, if you compare this with the preceding example you will find the work so far has been the same. That is because the 1 of 154 has not yet contributed to the answer—it has not yet paired with any part of the 273. But now we go on:

Third step:

U T ← (4)

U T ← (5)

U T ← (1)

0 0 0 2 7 3 × 1 5 4

••0 •4 2

4 gives $0\underline{8}+\underline{2}8$

5 gives $3\underline{5}+\underline{1}5$

1 gives $0\underline{3}+\underline{0}$

underlined figures add to 19 and the dot makes it 20

Fourth step:

```
U T←──────────┐
  U T←──────┐ │
    U T←──┐ │ │
000 273 × 1 5 4
───────
  ·2 ··0·4 2
───────
```

The 4 is not out of action yet—it still contributes the tens-digit of 2 times 4. But this happens to be zero, as 2 times 4 is 08. We get real contributions only from the 5: 10 plus 35 and from the 1: 07 plus 03.

Fifth step:

```
U T←─────────────────┐
    U  T←──────────┐ │
       U  T←─────┐ │ │
0 0 0  2  7 3  × 1 5 4
──────────────
  4 ·2 ··0 ·4 2
──────────────
```

the 4 is out of action, its UT *hits only zeroes*

```
4 gives   0+0
5 gives      0+10
1 gives         02+07
```

Last step:

```
 U T
000273  ×  1 5 4
──────
042042
```

The answer is 42,042. From this example it is evident how we proceed if the multiplier has four digits, or any number of digits.

We have considered multipliers of increasing length, first one digit, then two, then three, and of course this has resulted in a certain amount of repetition. Primarily this step-by-step procedure was chosen to ensure a clear presentation

—we wished to be sure the process would not be misunderstood. It should now be added that this step-by-step procedure was intended to do something more. It emphasizes the point that this is the correct way to go about practicing the method. Doing several dozen multiplications by single digit multipliers will make the two-digit multipliers seem as easy as they really are. Practicing with two-digit multipliers until the procedure is familiar will make longer multiplications easy. Altogether the amount of practice needed to master the whole process quite thoroughly is only a few hours. At the end of our practicing we shall be in possession of a new and interesting technique, and also a new speed in calculation. The degree of speed achieved is determined by the amount of practice we give to it; with sufficient practice the speed can become phenomenal.

SUMMARY

In brief, the two-finger method consists of these three features:

1. There is a process of forming pair-products, as the pair-product 7 is formed out of 5 3 times 7:

$$\begin{array}{ccl} \text{U} & \text{T} & \\ 5 & 3 & \times \quad 7 \\ \cline{1-2} 3\underline{5} & \underline{2}1 & \\ \multicolumn{2}{c}{5+2} & \\ \multicolumn{2}{c}{7} & \quad \textit{the pair-product} \end{array}$$

2. There is a way of multiplying any number by a single digit, using these pair-products:

$$\begin{array}{cccccl} & & & \text{U} & \text{T} & \\ 0 & 3 & 2 & 5 & 3 & \times \quad 7 \\ \cline{1-5} & & & & 1 & \quad \textit{the units of 3 times 7 is 21 is 1} \end{array}$$

```
      U T
0 3 2 5 3   ×  7
      7 1        7 is the pair-product of 5 3
                 times 7
```

until finally we reach

```
U T
0 3 2 5 3   ×  7
2 2 7 7 1
```

3. There is a way of extending this multiplication by a single digit to include multiplication by numbers of any length. This is done by forming several pair-products and adding them to find each digit of the answer. These several pair-products are "inner" and "outer" pairs, taken by moving inward from both ends toward the space between the two numbers that are being multiplied:

```
      U T
        U T
0 0 7 3       ×  5 4
      2
```

```
           U T
             U T
then     0 0 7 3    ×  5 4
            ˙4 2         this 14 is 8 plus 1 plus 5,
                         from 28, 12, 15
```

and so on, until the end:

```
U T
  U T
0 0 7 3  ×  5 4
3 9˙4 2
```

Remember, the U that begins the UT pattern in any problem—the U farthest to the left—is always placed directly above the position where the next figure of the answer will appear.

PRACTICE PROBLEMS

Say aloud the pair-product of the following pairs of numbers (answers are below):

1. 67 × 8 **3.** 94 × 2 **5.** 66 × 7
2. 56 × 4 **4.** 77 × 6 **6.** 59 × 7

Perform the following single-digit multiplications by using pair-products, one for each digit of the answer.

7. 56 × 4 **8.** 82 × 8 **9.** 3945 × 6

Perform the following multiplications by using pair-products and inner and outer pairs (two-products for each figure of the answer):

10. 95 × 62 **12.** 83 × 45 **14.** 43546 × 62
11. 38 × 66 **13.** 3456 × 86

Answers:

1. 13	**5.** 6	**9.** 23,670	**13.** 297,216
2. 2	**6.** 11	**10.** 5,890	**14.** 2,699,852
3. 8	**7.** 224	**11.** 2,508	
4. 6	**8.** 656	**12.** 3,735	

CHAPTER FOUR

Addition and the right answer

In the preceding chapters we developed methods of multiplication with emphasis on speed. At the same time, we paid attention to the need for accuracy and brought out the importance of checking our results.

In the problem of addition we have again these two factors, speed and accuracy. We shall develop in this chapter a method of addition which is faster than the method used by most people, and we shall also find a method for checking and double-checking the results. The emphasis, however, needs to be changed somewhat. In using conventional addition, the average man cannot always add a fairly long column of figures without making a mistake. Suppose he has a column of five-digit numbers to add. He must add five separate columns, and in one time out of the five he is likely to make a mistake, using the conventional method.

We shall learn how to check the work by individual columns, without repeating the addition. This has several advantages:

(1) *we save the labor of repeating all the work, while at the same time*

(2) *we locate the error, if there is one, in the column where it occurs (making it easy to correct), and*

(3) *we are certain to find the error, which is not necessarily the case if we repeat all the work.*

This last point is something that most people do not realize. Each one of us has his own weaknesses and his own kind of errors. We recognize this in spelling: a man who is otherwise good at spelling may spell "parallel" as "paralell," and another may spell "harass" with two r's. In arithmetic the same thing happens, though most of us are less aware of it. One person may have a perverse tendency to say that 8 times 7 is 54. If you ask him directly he will say "56," but in the middle of a long calculation it will slip out as "54." Other people have different quirks. That is one reason why repeating the calculation is a poor way to check it. Whatever mistake the calculator made the first time, it is probably his favorite error, and he will likely do it again when he checks by repetition.

"Natural" errors may trip us up if we persuade another person to check our work for us. These are the mistakes that come naturally to anyone in the same situation. Handwriting is the simplest example. If the one who wrote the figures makes the number 4 as it is in print, with the lines coming together at the top, then when he writes too fast it will round off into something very much like a 9. Anyone checking the work will read it as a 9, just as the first person did in making the original error.

Many other natural errors occur in various situations: continuing the repetition of a repeated pattern one time too many, reversals, and a variety of others. But in everyday life these natural errors, which are natural for everyone, are less important than our personal favorite errors, simply because we can seldom get anyone else to check our work for us.

All in all, we can make the broad statement that practically any other way of checking the work is better than repeating it. We have in the Trachtenberg system a particular way of checking. Also, we have a speed feature which will be new to many, perhaps most, people.

FINDING THE TOTAL

As in the conventional method of addition, we write the figures to be added in a column, and under the bottom figure we draw a line, so that the total will be under the column. When writing them remember that the mathematical rule for placing the numbers is to align the decimal points. That is, put all the decimal points directly under the first one. Adding 12.5, 271.65, and 3.01, we write:

```
  12.5
 271.65
   3.01
```

Very often we do not see any decimal point, as in the number 73, but then the decimal point is just after the number; 73. would be the complete form, which nobody bothers to write. So when no decimal points are seen in the numbers you are given, you line them up on the last figures, where the invisible decimal points are. That is what we have been doing in the conventional method of addition, and in our new method we set up the problem in the same way. Here is one set up correctly to work:

```
3 6 8 9
  7 5 8
9 6 6 7
1 0 6 4
6 4 9 8
  7 4 5
9 9 6 8
5 8 8 7
9 9 8 8
7 6 1 5
8 7 4 9
-------
```

The conventional method now has us adding the figures down the right-hand column, 9 plus 8 plus 7, and so on. We can do this if we wish in the new method, but it is not compulsory: you can begin working on any column. For the sake of novelty, we will start on the left-hand column. We add as we go down, but we use Trachtenberg's rule:

Never count higher than eleven.

That is, when the running total becomes greater than 11, we reduce it by 11 and go ahead with the reduced figure. As we do so, we make a small tick or check-mark beside the number that made our total higher than 11. In the example, we go down the left-hand column, repeated below on the left, making mental calculations as follows:

3		
9′	3 plus 9, 12:	this is more than 11, so we subtract 11 from 12. Make a tick and start adding again with 1.
1	1 plus 1, 2	
6	2 plus 6, 8	
9′	8 plus 9, 17:	make a tick and reduce 17 by 11. Say "6" and go on.
5′	6 plus 5, 11:	make a tick, say "zero," and go on.
9		
7′	9 plus 7, 16:	make a tick, say "5," and go on.
8′	5 plus 8, 13:	make a tick and write 2.

The final figure, 2, will be written under the column as the "running total."

Next we count the ticks that we have just made as we dropped 11's. How many were there in the example? Five. So we write 5 under the column as the "tick figure." The example now looks like this:

	3	6	8	9
		7	5	8
	9′	6	6	7
	1	0	6	4
	6	4	9	8
		7	4	5
	9′	9	6	8
	5′	8	8	7
	9	9	8	8
	7′	6	1	5
	8′	7	4	9
running total:	2			
ticks:	5			

The desired answer will be found from the running total and the tick number. But first we must do the same for the other columns. The result will be:

	3	6	8	9
		7′	5′	8′
	9′	6	6	7′
	1	0	6′	4
	6	4′	9′	8′
		7	4	5
	9′	9′	6′	8′
	5′	8′	8	7′
	9	9′	8′	8
	7′	6	1	5′
	8′	7′	4	9′
running total:	2	3	10	1
ticks:	5	6	5	7

Now we arrive at the final result by adding together the running total and the ticks in this way: we add the neighbor on the right in the bottom row of ticks. Like this:

```
 – 3 – –
 – 6 5 –
   14      3 plus 6 plus 5
```

In the example we have:

```
running total:  0 2 3 10 1
        ticks:  0 5 6  5 7
                         8     1 plus 7 is 8
                       ••2     10 plus 5 plus 7 is 22
                    •6         3 plus 6 plus 5 plus carry is 16
                  •4           2 plus 5 plus 6 plus carry is 14
                6              zero plus zero plus 5 plus carry is 6
```

This special kind of addition, adding in the neighbor at the lower right, is a regular feature of the method, and it will always be used.

In full, if the example we have just seen were to be done as actual work, the answer would look like this:

(columns of figures)

```
―――――――――――――
0 2 3 10 1
0 5 6  5 7
―――――――――――――
6 ˙4,˙6 ˙˙2 8
```

In adding these two rows we begin at the right-hand end and move to the left, as we are accustomed to doing in conventional addition. At the last step, we must imagine two zeroes as they are shown, one above the other, if we have not actually written them there. That is because there is something left to add, namely the leftmost digit in the tick row, the 5 in this case. This happens because we are not adding in the usual manner, we are adding "L-shaped." At the last step we look at these figures:

```
0
0 5 ˙
―――――
0 6    5 plus the carry is 6
```

The same procedure is to be followed in all cases.

A simple short-cut

To make our addition even easier, you will notice that when using the "elevens-rule" we cannot go over 19 when we are running down a column. The first or left-hand digit is therefore always a 1, when we go over 11. Consequently we do not need to perform a true subtraction when we "subtract 11." It will be sufficient to forget the first digit and reduce the other digit by 1: if we have 16 we think only of the 6 and reduce it to 5. So 16 becomes 5, as we make our tick. This sounds trivial, but it is not. In doing actual problems, the way you think can make the work twice as hard or twice as easy.

For a short and simple example, suppose you add up a problem in dollars and cents:

```
  .8 9
  .2 3
  .9 6        Stop at 11
 1.0 4           and
  .3 9       make a tick!
  .2 5
 -----
```

Did you remember to add in the lower right-hand neighbor, when you added the two bottom rows? If you did, you must have arrived at the right answer. It is 3.76. To be specific, the bottom rows are:

```
 1.2 3
 0.2 3
 -----
```

and these do actually add up to 3.76 if you add in the lower right-hand neighbors in the bottom row, the 0.23.

Example 1:

```
                   5 4 7 7
                   9 6 6 5
                   2 7 4 6
                   8 3 5 6
                   7 4 9 9
                   5 1 6 2
                   6 8 7 5
                   -------
running totals:    9 0 0 7
         ticks:    3 3 4 3
                   -------
         total:  4˙5 7 8˙0
```

Example 2:

	1 6 . 3 9
	5 0 7 . 2 6
	1 9 5 . 0 0
	7 8 . 3 7
	6 4 . 2 7
	4 . 7 5
	8 8 . 4 7
	2 8 6 . 5 5
running totals:	8 6 4 . 4 2
ticks:	0 3 4 . 2 4
total:	1 ·2 ·4 ·1 .·0 6

Here are a few problems for practice, with the correct answers following. It will be easy to make up similar problems for yourself:

Problem 1:

4 6 9
7 4 2
3 2 5
9 6 2
5 2 7
6 2 3
2 1 3

Problem 2:

6 1 5 9 8
5 0 4 2 3
7 2 4 6
7 4 4
4 2
9 3 5 7
2 1

Problem 3:

```
  1.25
  3.06
  7.58
   .98
 38.50
 59.50
  9.75
  2.98
 12.25
 14.85
 45.00
 25.75
______
```

Problem 4:

```
166.15
 35.94
 34.13
705.75
422.50
  2.99
 16.77
522.35
875.88
 27.66
 55.18
149.75
______
```

The answers below were obtained by applying L-shaped addition to the following running totals of reduced digits and tick numbers:

(1)

```
0 3 1 9
0 3 2 2
_______
3 8 6 ·1
```

(2)

```
0 0 6 10 8 9
0 1 1  1 2 2
____________
1 2 9 ·4 ·3 ·1
```

(3)

0	5	0	4	0
0	1	5	5	5
2	·2	·1	·4	5

(4)

0	4	2	3	9	10
0	2	3	5	5	5
3	·0	·1	·5	··0	·5

NOTE. From time to time, various persons have used a procedure similar to our "elevens-rule:" they subtracted 10 and made a tick whenever the running-total exceeded 10. This is a good idea as far as it goes. However, we prefer the "elevens-rule" because it goes further and can give us a special check and double-check which will be detailed later in this chapter.

CHECKING THE ANSWER

To summarize in a few words what we have done, let us look again at one of our examples, but now with most of the figures only suggested by dots:

column of figures:		3	6	8	9
		·	·	·	·
		·	·	·	·
		8	7	4	9
working-table:		2	3	10	1
		5	6	5	7
answer:	6	4	6	2	8

Summarizing in a few words, we used the column of figures to make up the working-table, and we used the working-table to find the answer. The working-table, as you see, consists of the running totals and the tick numbers that we have been using in every example.

Now to check our work we use these three items, the column of figures, the working-table, and the answer. We calculate a check-figure from each, and we compare these check-figures with one another to see if they agree. If they do, the work is correct. If they do not agree, there is something wrong somewhere. With this method of checking, the "somewhere" quickly becomes specific. We will determine in the check, without repeating the addition, which column of figures was added incorrectly.

Since we have three items to check, the act of checking consists of three steps. First, let us describe the three steps, and then we shall do each of them in detail as applied to our example:

First step: We find a check-figure for each column of figures.

Second step: We find a check-figure for the working-table.

Third step: We find a check-figure for the answer (or total).

First step: Taking in turn each column of digits, we find the nines-remainders. This is the same as the reduced digit-sums that we saw on page 78. To find the nines-remainders we strike out, or underline, all 9's and all combinations of digits that add to 9 or to multiples of 9; then we add up only what remains. This will often be a two-digit figure. If it is, we reduce it to a single digit, by adding the two digits together. The resulting single figure is the check-figure for that column of digits. For instance, we find the check-figure for the left-hand column of digits in one of our previous examples in this way:

~~3~~	6	8	9
	7	5	8
~~9~~	6	6	7
~~1~~	0	6	4
~~6~~	4	9	8
	7	4	5
~~9~~	9	6	8
5	8	8	7
~~9~~	9	8	8
7	6	1	5
~~8~~	7	4	9
12			
3			

Besides the 9's in the first column, we have struck out 3 and 6, because they add to 9. Likewise we struck out 1 and 8.

Add the remaining numbers: 5 plus 7 is 12. To use in checking we shall reduce this to a single figure by adding across. This gives 1 plus 2 equals 3. So the running check number for the first column is 3.

The other three columns are treated in the same way. If you happen to notice three figures in a column which add to 9, strike out all three of them. But if you should overlook such a set of three figures there will be no harm done. It is always true that you get the same final *single* figure, after adding across as we added the 1 plus 2 of 12, no matter how many opportunities for striking out you may have missed. Adding the digits across compensates for the missed opportunities, and the only loss is the loss of a small amount of mental energy.

Try doing the other three columns yourself. When you finish the result should look like this, with the cast-out figures underlined instead of struck out:

$$
\begin{array}{rccccc}
 & & \underline{3} & \underline{6} & \underline{8} & \underline{9} \\
 & & & 7' & \underline{5'} & 8' \\
 & & \underline{9'} & \underline{6} & \underline{6} & 7' \\
 & & \underline{1} & \underline{0} & \underline{6'} & \underline{4} \\
 & & \underline{6} & 4' & \underline{9'} & 8' \\
 & & & 7 & \underline{4} & \underline{5} \\
 & & \underline{9'} & \underline{9'} & \underline{6'} & 8' \\
 & & 5' & 8' & 8 & 7' \\
 & & \underline{9} & \underline{9'} & 8' & 8 \\
 & & 7' & \underline{6} & \underline{1} & 5' \\
 & & \underline{8'} & 7' & 4 & \underline{9'} \\
\hline
\text{running totals:} & & 2 & 3 & 10 & 1 \\
\text{ticks:} & & 5 & 6 & 5 & 7 \\
\hline
\text{total:} & 6 & 4 & 6 & 2 & 8
\end{array}
$$

CHECK:

nines-remainders 3 6 2 6

Reducing to a single figure in this way need not be left to the end. It is better to do it as you go along. (In the second and third columns from the left, note that the three cast-out 6's total 18, a multiple of 9.) In the second column from the left, you may if you wish add the figures not underlined, obtaining 33, and reduce 33 to 6, as shown. It is better, however, to do it as you go along because it is easier. Going down the column, and ignoring all underlined figures, we have 7 plus 4, 11, "is 2" (because 1 plus 1 is 2), plus 7 is 9, plus 8, 17, "is 8" (because 1 plus 7 is 8), plus 7, 15, "is 6" (because 1 plus 5 is 6). This 6 is the check-out figure, and of course it agrees with what we had before. The point is merely that it is easier to work with the small figures that we get if we reduce as we go along, instead of adding up to 33 and reducing the 33 to 6. We shall go into this again a little later.

Second step: The purpose of this step is to check the working-table, which in the example was this:

running totals:	2	3	10	1
ticks:	5	6	5	7

We find check-figures for this table by repeating the second line and adding:

running totals:	2	3	10	1
ticks:	5	6	5	7
repeat ticks:	5	6	5	7
Adding, we have:	12	15	20	15
Reduced, we have:	3	6	2	6

Compare these last figures with what we found in the First step. We found four single-figure numbers, 3, 6, 2, 6, corresponding to the four columns of the addition. In the Second step we have just found four single-figure numbers, 3, 6, 2, 6. These agree exactly with the check-figures from the First step, so the work is correct.

Suppose it were to happen that the two sets of four figures did not agree. Suppose, for instance, that the First step gave 3, 6, 7, 6, as compared with the 3, 6, 2, 6 from the Second step. There is disagreement in the third figure from the left. Then we know that the third column from the left was added incorrectly, but the other three columns are correct. The error can be found by looking at column three only.

Third step: This step obtains a check-figure from the answer. In the example the answer, or total, was 64,628. The check-figure is the digit-sum of this: 6 plus 4 plus 6 plus 2 plus 8 is 26, which reduces to 8.

What shall we compare this with, to check? We compare it with the two sets of four single-figure numbers that we found in the First and Second steps: the figures 3, 6, 2, 6.

Adding these, we have 17, which reduces to 8. The total of 64,628 also gave the reduced digit-total of 8. The 8's agree, so everything checks.

This procedure is used in every addition. In practice, of course, we do not bother to write down all the figures that were shown in the explanation above. In checking the working-table particularly, it is not necessary to rewrite the table, with the bottom line repeated. The repetition of the bottom line can be done mentally by simply adding in the bottom figure twice. So the example that was used just now in the explanation of the method would look like this is an ordinary addition:

$$
\begin{array}{rccccccl}
 & & \underline{3} & \underline{6} & \underline{8} & \underline{9} & & \\
 & & & 7' & \underline{5'} & 8' & & \\
 & & \underline{9'} & \underline{6} & \underline{6} & 7' & & \\
 & & \underline{1} & \underline{0} & \underline{6'} & \underline{4} & & \\
 & & \underline{6} & 4' & \underline{9'} & 8' & & \\
 & & & 7 & \underline{4} & \underline{5} & & \\
 & & \underline{9'} & \underline{9'} & \underline{6'} & 8' & & \\
 & & \underline{5'} & \underline{8'} & 8 & 7' & & \\
 & & \underline{9} & \underline{9'} & 8' & 8 & & \\
 & & 7' & \underline{6} & \underline{1} & 5' & & \\
 & & \underline{8'} & 7' & 4 & \underline{9'} & & \\
\hline
\text{running totals:} & & 2 & 3 & 10 & 1 & & \\
\text{ticks:} & & 5 & 6 & 5 & 7 & & \\
\hline
\text{answer:} & 6 & 4 & 6 & 2 & 8 & & \text{TOTAL}
\end{array}
$$

CHECK:

$$
\begin{array}{rcccccll}
\text{columns:} & & 3 & 6 & 2 & 6 & & \text{(nines-remainders of columns)} \\
\text{working-table:} & & 3 & 6 & 2 & 6 & = & 8 \\
\text{answer:} & 6 & 4 & 6 & 2 & 8 & = & 8
\end{array}
$$

In practical work we would also omit all the words and use only figures. When you do an addition yourself, you know that you meant the line 2, 3, 10, 1 to be the running totals, and so on. We have labeled them here only to avoid any possible misunderstanding.

Here is another example, set up just a little differently, which you may find more convenient. The difference is that the tick line of the working-table is repeated in writing, but the table itself is not rewritten: we repeat the tick line under the answer. Then we add up the three lines of the augmented working-table by jumping over the answer:

		.	8	$\underline{9}$	
		.	2	$\underline{3'}$	
		.	$\underline{9'}$	$\underline{6}$	
	1	.	0	$\underline{4'}$	
		.	$3'$	$\underline{9}$	
		.	2	$\underline{5'}$	
running totals:	1	.	2	3	
ticks:	0	.	2	3	
total:	3	.	7	6	the answer is 3.76
repeating ticks:	0	.	2	3	
check-figures:	1	.	6	9	
nines-remainders of columns:	1	.	6	0	

This checks, of course. In nines-remainders, a 9 and a zero are equivalent. The 1.69 was obtained by adding three figures downward: 1 plus zero plus zero gave the 1; 2 plus 2 plus 2 gave the 6; and 3 plus 3 plus 3 gave the 9.

Final part of the check: we check the answer, 3.76, by finding its reduced digit-sum. This is 7. The digit-sum of 1.69 is also 7, so this checks also. There is no error.

GENERAL METHOD OF CHECKING

In all kinds of calculations it is important to have some way of checking our work other than repeating it. Whether we are doing an addition, a subtraction, a division, squaring a number, taking a square root, or any combination of these, we need a good method of checking. Such a method exists and *applies to all kinds of calculations.* In fact, there are two such methods which are only slightly different, and for the sake of completeness we shall demonstrate both. We give the digit-sum method first, as the principal method, and the eleven-remainders as an alternate or optional method.

Digit-sum method

This may also be called the nines-remainder method. It is an old idea, adopted into the Trachtenberg system. You saw it appearing as part of the check on addition. As you probably remember, the concept of a digit-sum consists of this:

(1) You find the digit-sum of a number by "adding across" the number. For instance, the digit-sum of the number 5,012 is 5 plus 0 plus 1 plus 2 is 8.

(2) You always *reduce to a single figure,* if it is not already a single figure. For instance, the digit-sum of 5,012,431 is 5 plus 0 plus 1 plus 2 plus 4 plus 3 plus 1 is 7 (16, or 1 plus 6 is 7).

(3) In "adding across" a number you drop out 9's. In fact, if you happen to notice two digits that add up to 9, like 1 and 8, you ignore both of them. So the digit-sum of 9,099,991 is 1, at a glance. You don't bother to add up the 9's. (But if you did, you would *still* end up with the same 1, after you reduced to a single figure. Try it if you don't believe it!)

(4) Because "nines don't count" in this process, as we saw in (3), a digit-sum of 9 is the same as a digit-sum of zero. The digit-sum of 513, for instance, is zero. In certain cases it saves work if you remember this.

For example, what is the digit-sum of 918,273,645? You should be able to do it in about three seconds, without any hard thinking. The result is zero. That is because we ignore the 9; then we ignore pairs of numbers that add up to 9 and in this example each adjacent pair of numbers, after the first 9 adds up to 9. Everything drops out, and we end up with zero.

What is the digit-sum of 234,162? (*Hint: ignore any three digits that add up to 9.*) Again we have zero.

Usually, of course, the number we are looking at will contain some digits that do not add up to 9. Whatever they do add up to is the digit-sum, after it is reduced to a single figure. So the digit-sum of 903,617 is 8. The 9 and the zero are ignored, 3 plus 6 is 9, and we are left with 1 plus 7 equals 8.

Work-saver: When you are "adding across" a number, as your running total reaches two digits you add these two together, and go ahead with this single figure as your new running total.

For instance: find the digit-sum of 7,288,476,568. Say 7 plus 2 is 9; forget it. Then 8 plus 8 is 16, a two-figure number. Reduce this 16 to a single figure: 1 plus 6 is 7. Go ahead with this 7: 7 plus 4 is 11, two figures, so we reduce it to a single figure, 1 plus 1 is 2. Go ahead with the 2: 2 plus 7 is 9, "is zero," and we start over. Then 6 plus 5 is 11, "is 2," and 2 plus 6 is 8. Then 8 plus 8 is 16, "is 7." So the digit-sum of this long number is 7.

Decimals work the same way exactly. We simply pay no attention to the decimal point. The digit-sum of 5.111, for instance, is 8.

Explanation: It is not necessary in a practical sense to understand why the method works, but I think you will find the explanation interesting. The basic fact is this: the numbers that we have been calculating, these reduced digit-sums, are precisely the remainders that you would get if you divide each number by 9. For instance, take 32. Divide it by 9: 9 times 3 is 27, and we have a remainder of 5. Take a longer number, say 281, and divide it by 9; you find a quotient of 31, and there is a remainder of 2. But in the first case, as you have no doubt noticed, 32 has a digit-sum of 5, equal to the remainder 5, and in the second case the digit-sum of 281 is 11, which reduces to 2. In every case our digit-sum, after it is reduced to a single figure, will be equal to the remainder after dividing by nine.

Application to checking

How are we going to use these digit-sums to check our calculations? It seems to work differently in different cases, but really we need only remember one underlying principle:

> **Basic rule:** Whatever you do to the numbers, you also do to their digit-sums; then the result that you get from the digit-sums of the numbers must be equal to the digit-sum of the answer.

For instance: suppose that the operation involved is multiplication, and we are multiplying 92 by 12. The product is 1,104. We can write it in parallel lines:

the numbers:	9 2	×	1 2	=	1 1 0 4
the digit-sums:	2	×	3	=	6
	(1+1)		(1+2)		(1+1+0+4)

The digit-sum 2 comes from 92 or 9 plus 2. This gives 11, which we reduce to a single figure: 1 plus 1 is 2. (Or, we may simply ignore the 9.) The point is that we get the 6 on the right-hand side in two ways. One way is from the left-hand side: 2 times 3 is 6. The other way is from the answer, 1,104: 1 plus 1 plus 0 plus 4 is 6. Of course, we say 6 equals 6, and the work checks. Our 1,104 is right.

The method works equally well in addition:

the numbers:	1 5	+	1 2	+	2 0	=	4 7
the digit-sums:	6	+	3	+	2	=	1 1 (4+7=11)
which reduces to:			2			=	2

In the first example we were *multiplying* the given numbers, 92 times 12, so we had to *multiply* the digit-sums, 2 times 3. In the second example it was different. We were *adding* the given numbers, 15 plus 12 plus 20, so we had to *add* their digit-sums, 6 plus 3 plus 2. We always carry out a parallel calculation, using the digit-sums instead of the numbers.

Of course, the given numbers will frequently be quite large, sometimes in the millions. Their digit-sums will always be small; in fact, they reduce to a single figure. Consequently, this check requires only a small amount of calculation and gives us a valuable verification of the work.

Check this double multiplication:

$$322 \times 28.1 \times 12.4 = 112{,}197.68$$

Ignore the decimal points in checking. To the left of the equals sign we have these digit-sums:

7	×	2	×	7	
		(14)			*Mentally—don't write*
		(5	×	7)	*Mentally—don't write*
			(35)		*Mentally again!*
			8		The digit-sum

On the right-hand side of the equals sign we have the answer, 112,197.68. The digit-sum of this, by adding across, is 8. Then 8 equals 8, and the work is correct.

In simple cases, division works in exactly the same way. This is an example of it:

numbers:	1 3 2	÷	1 1	=	1 2
digit-sums:	6	÷	2	=	3

Thus, the digit-sum of the answer is 3 (1 plus 2), and we also get 3 by dividing 6 by 2. So the work checks.

But more often, division proves to be a little more complicated than this because it often does not "come out even." We shall go into this later in the chapter on division. In the meantime, it will be sufficient to note this fact:

Division can be checked by *multiplying* the appropriate digit-sums.

For instance, in the example just above, we can multiply the digit-sum of the quotient by the digit-sum of the divisor, that is, 3 times 2. This gives 6, the digit-sum of the dividend, and the work checks.

The elevens method

Instead of the digit-sum method, we can use this alternate method either as a double check, if such a thing is desired, or simply for the sake of variety. This is the elevens-remainder method. But we do not divide anything by 11. Just as the regular digit-sum is the remainder that we would get after dividing by 9, we now find the remainder that we would have after dividing by 11, and we find it in a way somewhat similar to the digit-sums. This is the method:

FIRST CASE: TWO-DIGIT NUMBERS.

To find the elevens-remainder of a two-digit number, like 48, we subtract the tens-digit from the units-digit: for 48 we have 8 minus 4 is 4. The elevens-remainder of 48 is 4. This is what we would have found if we had actually divided 48 by 11.

Sometimes we can't subtract because the tens-digit is larger than the units-digit, as it is in 86, for instance. In that case, we *make* the units-digit big enough by adding 11 to it. For 86 we have 6 plus 11 is 17, minus 8 is 9. For 52 the elevens-remainder would be 2 minus 5; make it 2 plus 11 minus 5 is 8.

SECOND CASE: ALL NUMBERS LONGER THAN TWO DIGITS.

The method here is to use *every second digit.* That is to say, we start at the right-hand end of the number, and work back to the left adding up every second digit, and then we pick up the ones we have skipped and *subtract* them. Take 943,021,758. Start at the right-hand end, the 8, and work back to the left adding up every second digit:

$$8+7+2+3+9 \quad = \quad 29$$

Then go back to the next-to-last digit, the 5 of the long number, and again pick up every second number:

$$5+1+0+4 \quad = \quad 10$$

Then subtract:

$$29 \quad - \quad 10 \quad = \quad 19$$

This 19 still has to be reduced, just as we had to reduce our digit-sums to a single figure before. In this method we reduce it by subtracting the tens-digit from the units-digit:

$$9 - 1 = 8$$

We had 29 minus 10 in this example. Suppose in another case we had something like 29 minus 35, so that we couldn't subtract; what would we do then? We would add 11 to the smaller number to bring it up to where we could subtract; here 29 minus 35 would become 40 minus 35 equals 5.

Do these numbers seem uncomfortably large, the 29 and the 35? They can easily be avoided by using little short-cuts. Here is one, similar to one that we had before: after you find the first number, like the 29, do *not* set it aside to find the 35. Instead, you will now work down from the 29, or whatever number it is. That is, after you find the total of every second number, go back to the next-to-last number and subtract it from the total. Continue to the left through the number, subtracting every second digit from this new starting-point. (These are the numbers we skipped to reach the first total.) It amounts to subtracting the second total a little piece at a time. Take 2,368,094. Start at the end with the 4, and add the underlined figures:

$$\underline{2},3\ \underline{6}\ 8,\underline{0}\ 9\ \underline{4}$$

Add 4 plus zero is 4, plus 6 is 10, plus 2 is 12. Then go back using the other figures,

$$2,\underline{3}\ 6\ \underline{8},0\ \underline{9}\ 4$$

and subtract by going down from the 12 that we found just now: 12 minus 9 is 3; then minus-8 won't go because 8 is too large for the 3, so we increase 3 by 11 and say: 3 plus 11 is 14 minus 8 is 6, minus 3 is 3. The elevens-remainder is 3. If you prefer you could also go back, after adding up to the 12, and subtract the other figures from left to right: 12 minus 3 is 9, minus 8 is 1; 1 plus 11 is 12; 12 minus 9 is 3. The elevens-remainder of this number will always be 3, no matter what short-cuts you use.

A different short-cut, very effective, is to go across the number using adjacent pairs of figures. In each pair we subtract one figure from the other, because one of them is an "even" figure and the other is an "odd" one (in order of position, of course). For instance, take 4,693,260,817. Write the number out and group the digits into pairs, as we show by underlining:

	4 6	9 3	2 6	0 8	1 7
Subtracting:	2	5	4	8	6

Explanation: the upper line is the given number with the pairs of figures indicated, and the lower line is the set of elevens-remainders that we get from these pairs. We get each one from its pair in the usual way, by subtracting the tens-digit from the units-digit. Of course these are only temporarily treated as units- or tens-digits; we think of the 4 6 at the beginning as if it were 46, for the purpose of doing this little calculation. Going across the whole number from left to right, the 4 6 gives us 6 minus 4 is 2, then 3 minus 9 becomes 3 plus 11 is 14; 14 minus 9 is 5; then 6 minus 2 is 4; then 8 minus zero is 8; then 7 minus 1 is 6. This accounts for the row of figures under the given number, one figure to each pair.

Now we add these figures that we have just found: 2 plus 5 is 7, plus 4 is 11, "is zero," we say, because in elevens-remainders an 11 is equivalent to zero; then 8 plus 6 is 14, "is 3," we say, reducing by 11 as always in the elevens-remainder method. The result is 3.

Applications: We apply the elevens-remainders to calculations as a check, in the same manner as we did with the nines-remainders previously. The principle involved is the same as before:

> Whatever operation we perform on the given numbers, we do the same operation on the elevens-remainders. Then the result of operating on the elevens-remainders must be the same as the elevens-remainder of the answer, if the answer is correct.

For example, let us check a multiplication that we performed in a previous chapter. We arrived at the result of 302 times 114 equals 34,428. Let us write these numbers out, each with its elevens-remainder below it:

$$\underset{(5)}{3\ 0\ 2} \quad \times \quad \underset{(4)}{1\ 1\ 4} \quad = \quad \underset{(9)}{3\ 4\ 4\ 2\ 8}$$

Multiply the 5 by the 4: the result should be the 9 on the right-hand side of the equation, if the multiplication is correct. This means, of course, they must be equal in the sense of the remainders, after we drop out eleven if necessary to reduce to a single figure. Is this the case here? Yes. Because 5 times 4 is 20; remove 11, the 20 is reduced to 9. We have the equation of elevens-remainders 9 equals 9, parallel to the original multiplication.

Here are two more examples. See whether you can check them yourself by using this method:

(1) $5\ 2\ 7\ 3 \times 5\ 4 = 2\ 8\ 4\ 7\ 4\ 2$

(2) $2\ 7\ 3 \times 1\ 5\ 4 = 4\ 2\ 0\ 4\ 2$

You should find that both are correct. In (1) we have the elevens-remainders, 4 times 10 equated to 7: that is, 40 has the same elevens-remainder as 7. Reducing 40 by subtracting the 4 from 0 (increased by 11), we find this is true.

In (2) we have the elevens-remainders 9 times zero on the left, and zero on the right; 9 times zero is zero, so the work checks.

CHAPTER FIVE

Division—speed and accuracy

It was the first day of school in a large American university. In one of the rooms thirty students of the first-year algebra class had gathered to hear a lecture from the head of the mathematics department. He had taken the assignment himself for a good reason—he wanted to see that the students had a solid foundation for their later work. You can't build on a poor foundation, and nowadays a poor foundation is what you have to work with most of the time.

So he did the first thing that must be done in teaching fundamentals. He convinced the students that they needed teaching, by deflating their over-confidence. He deflated them in this way: he gave them a rather long long-division problem to work out. On the blackboard he wrote a long number, something like 7,531,264, and he asked them to divide it by something like 9,798. They all pitched right in and after a while even the slow ones were finished.

Then he took up their worksheets and looked at the answers. Out of the thirty students he had twenty-five different answers, one right and twenty-four wrong. Six of the thirty had done the work correctly, but twenty-four had made at least one mistake somewhere along the line.

Why was this? Remember, these were college students. They had all learned the method in grammar school, had studied further mathematics in high school, and had passed all their courses. The same test applied to the general population would have given even worse results. The explanation is that we are not taught to be sure of getting the right answers. We are not made to appreciate the fact that a problem is not finished until we have the *right* answer, not just an answer. In fact, a problem is not really finished until we have proved that we have the right answer.

In the last chapter we emphasized the importance of systematic checking. Now that we are coming to division, it is even more important to have such a habit than it was before—in multiplication and addition.

To take care of this need we offer a choice between two methods of long division. Both are different from conventional long division.

The first is the "simple" method, intended for people who are more or less non-mathematical—that is, they are in lines of work which require little mathematics, or they have only a limited interest in doing mathematics for its own sake. An appropriate method of long division for such persons would be one that is easily remembered and is as nearly foolproof as possible in the matter of arriving at the right answer.

The other one is the "fast" method. Anyone who has a liking for figures will enjoy this. There is just enough to it to make it stimulating to one with some aptitude for mathematics and, once learned, it is much easier to do than the conventional method. Also, when it has been thoroughly mastered, it becomes really impressive to watch. The answer to a long division can be written down immediately, without any intermediate calculation.

THE SIMPLE METHOD OF DIVISION

This requires no aptitude for mathematics. We need only be able to add two numbers together and to do simple subtraction.

We are going to divide 27,483,624 by 62. The setup we use is similar to the one that most people use:

```
6 2     2 7 4 8 3 6 2 4     answer
```

We call the 62 the "divisor," which is a very natural name. As we do the work, this 62 becomes the top figure of a column of figures. We get this column by adding 62 repeatedly, ten times to be exact:

```
  6 2     2 7 4 8 3 6 2 4     answer
  6 2
 -----
1 2 4
  6 2
 -----
1 8 6

and so on.
```

On the left of the divisor column we are going to set up a column of digit-sum check-figures that will look like this:

```
Check            Divisor
column           column
 8                 6 2
 8                 6 2
----             -------
(16) → 7         1 2 4
       8           6 2
    -----        -------
 6 ← (15)        1 8 6

                 and so on.
```

Now let us look at how we find the check-figures. At each step as we add in another 62 in the divisor column, so in the check column we add in another 8. This is because 8 is the digit-sum of 62 (6 plus 2 is 8). When we go into two-figure numbers (as we did in this example as soon as we added 8 plus 8 is 16), we *cut it down immediately* to a single figure, by merely adding the two figures together. Here we had a 16, so we changed it to a 7 (1 plus 6 is 7). Then we go ahead with the 7. Add another 8 at the next step. You have 7 plus 8 is 15. But this is a two-figure number, so we cut it down to a single figure, 6 (1 plus 5). And so on every time.

What do you do with these check-figures? You use each one just as soon as you find it. After the first addition, you have the 16 which reduces to the 7. Notice that this is directly to the left of the first of the main additions, the 124. So you compare the 7 with the 124. Add across the 124, you have 1 plus 2 plus 4 is 7. This agrees with the 7 that you already have. So this line is correct. Then you add another 62 to the 124, and you get 186. In the check column on the left you add another 8 and you get 15, as we saw above. This 15 reduces to 6, which is the check-figure for the 186. Does it really check out? Add up 186, across; 1 plus 8 is 9, which we drop out and forget (*always drop nines* in digit-sum checking!) and we have left the 6. So 6 equals 6, and the 186 line checks. At each step, you add another 62 in the divisor column and another 8 in the check column. As soon as you have done that, you compare the new figure in the divisor column with the new figure in the check column—"comparing" meaning that you add across the new divisor total and see if it agrees with your new check-figure.

Obviously, if you make this check at each step as you go along, you will discover any error in your addition as soon as it occurs. This keeps everything on the right road all the time.

How far should you go in this way? Ten times. To be exact,

the number 62 should be written in the column ten times, so it has been added in nine times. The tenth result must be 620. This is simply the original divisor, whatever it was (62 here, but whatever it may be), with a zero added at the end. Putting a zero at the end of a number multiplies it by 10. Consequently this tenth number must be the original divisor with a zero after it. Here is the full column with all the checks:

Check column	*Divisor column*		*"long number" (dividend)*	*The answer*
8	6 2	**(1)**	2 7 4 8 3 6 2 4	will appear here
8	6 2			
(16) → 7	1 2 4	**(2)**		
8	6 2			
6 ← (15)	1 8 6	**(3)**		
8	6 2			
(14) → 5	2 4 8	**(4)**		
8	6 2			
4 ← (13)	3 1 0	**(5)**		
8	6 2			
(12) → 3	3 7 2	**(6)**		
8	6 2			
2 ← (11)	4 3 4	**(7)**		
8	6 2			
(10) → 1	4 9 6	**(8)**		
8	6 2			
0 ← (9)	5 5 8	**(9)**		
8	6 2			
8	6 2 0			

This 620 equals 62 times 10. It checks.

Once the divisor column has been set up and checked for correctness by the digit-sum figures the check column is no longer needed and may be stricken out.

Notice that we have also labeled the steps as we went along, with the bold numbers in parentheses. Each one of these label-numbers tells us what 62 has been multiplied by. Beside the **(2),** for instance, is 124. Now 124 is 62 times 2. The label-numbers identify the various multiples of 62. For instance, 434 is a multiple of 62, because 434 is 7 times 62. So you see 434 in our divisor column, and alongside of it is the indicator **(7),** to identify it as the 7-times multiple of 62.

The divisor column that we now have eliminates the need for any multiplication, and it is in multiplying that most errors are made. The rest of the method is in this rule:

Subtract repeatedly from the dividend, the largest number that you can use in the divisor column.

You start subtracting at the left-hand end of the long number (the dividend)—as in conventional division. At each step you look down the divisor column and find the biggest number there which is not *too* big, that is, not too big to subtract. Look at the example, 27483624. If you try to use only the first two figures, you have 27. Look down the divisor column. There is no number there which is smaller than 27. So take the first three figures of the long number, 274. Now look down the divisor column. What do you see that is smaller than 274? (You want it to be smaller, because you are going to subtract it from the 274.) Well, 62 is smaller than 274, and so is 124, and 186, and 248. The rest are larger than 274. So the largest number that we can subtract is 248. This number leads us to our next rule:

The label-number, or multiple number, of the number that we subtract is the next figure of the answer.

Here the label-number of 248 is **(4).** That means that 4 is the first figure of the answer:

Division column		*Dividend*	*Answer*
6 2	**(1)**	2 7 4 8 3 6 2 4	4
1 2 4	**(2)**	2 4 8	
1 8 6	**(3)**	2 6 8	
2 4 8	**(4)**		

and so on.

After writing this figure of the answer, you write the number that you wish to subtract under the long number (the dividend) and you perform the subtraction as shown above. You get 26 by subtracting. Then you carry down the next digit of the dividend. This is what you are accustomed to doing in conventional long division.

Now repeat the same process, using this new number under the long number. Here it is 268. You look down the divisor and find the largest number that is not *too* large. Here the number that you find in the divisor column is 248. Subtract it, after writing its label-number **(4)** as part of the answer:

6 2	**(1)**	2 7 4 8 3 6 2 4	4 4
1 2 4	**(2)**	2 4 8	
1 8 6	**(3)**	2 6 8	
2 4 8	**(4)**	2 4 8	
and so on		2 0 3	

The example fully worked out looks like this:

```
   6 2    (1)   2 7 4 8 3 6 2 4     4 4 3 2 8 4
 1 2 4    (2)   2 4 8
                -----
 1 8 6    (3)     2 6 8
 2 4 8    (4)     2 4 8
                  -----
 3 1 0    (5)       2 0 3
 3 7 2    (6)       1 8 6
                    -----
 4 3 4    (7)         1 7 6
 4 9 6    (8)         1 2 4
                      -----
 5 5 8    (9)           5 2 2
 6 2 0   (10)           4 9 6
                        -----
                          2 6 4
                          2 4 8
                          -----
                            1 6    the remainder
```

So the answer is 443,284, and there is a remainder of 16.

In practice, you will probably find the following point a help in saving trouble. In making up the divisor table, we must add in the divisor repeatedly, but this does not necessarily mean that we must write out the divisor many times. It is easy enough to look back at the head of the column where the divisor appears, and thus add the divisor to the number we have last obtained. In this way, we would have the work looking as follows:

```
               364,095 ÷ 465
               -------------
 465    (1)    3 6 4 0 9 5          7 8 3  answer
 930    (2)    3 2 5 5
               -------
1395    (3)      3 8 5 9
1860    (4)      3 7 2 0
                 -------
2325    (5)        1 3 9 5
2790    (6)        1 3 9 5
                   -------
```

3255	**(7)**	
3720	**(8)**	
4185	**(9)**	
4650	**(10)**	*check*

Here are some practice examples, which you may find interesting to try:

1. 73,458 ÷ 53 **2.** 90,839 ÷ 133
3. 23,525,418 ÷ 3,066 **4.** 21,101,456,770 ÷ 326

Answers: **1.** 1,386; **2.** 683; **3.** 7,673; **4.** 64,728,395

It might sometimes happen, though it is unlikely, that the person doing the work might be so careless as to choose the wrong figure out of the divisor column. This is unlikely because all we have to do is see which number is the biggest of those we can use. The biggest allowable figure in the divisor column will be the last one of those that are small enough to subtract, and all the figures following it will be too large. But suppose that someone did make such an error. There would still be no real harm done, because he would notice immediately that there was an error at that step:

(1) *If he took a number that was larger than the correct one, he would be unable to subtract it.*
(2) *If he took a number that was too small, then at the next step he would discover that the next "digit" of the answer was 10, which is not a digit.*

To check the subtraction, it is convenient and adequate to check them all at once by checking the answer itself. We do this by the following method:

(1) *Subtract the remainder from the dividend and take the digit-sum of what you get. In our example on page 140, we were left with a remainder of 16:*

dividend	27483624	
remainder	—16	
	27483608	= 2 *digit-sum*

(2) *Multiply the digit-sum of the answer by the digit-sum of the divisor:*

answer	443284 =	7	*digit-sum*
divisor	62 =	×8	*digit-sum*
		56 = 2	*digit-sum*

(3) *Compare the results and if they agree the work is correct. The answer is 2 in both cases, so the work is correct.*

THE FAST METHOD OF DIVISION

Do you remember, back when we were doing multiplication, we had what we called the "units-and-tens" method? We are now going to borrow one idea from that method and use it in division, but we shall add a new wrinkle to it. To refresh your memory we repeat what we had before. Taking a pair of digits, like 4 3, and a single multiplier like 6, we multiplied in a special way and obtained a single figure, namely 5:

	U	T	
	4	3 ×	6
the work	$2\underline{4}$ +	$\underline{1}8$	
the result	5		

The 24 is 4 times 6. The 18 is 3 times 6. Because the 4 of 43 has a U above it (U for units), we use only the units-digit of the 24, that it, we use the 4. Because the 3 of 43 has a T above it, for "tens", we use the 1, which is the tens-digit of 18. Then we *add* this 4 and this 1: 24 plus 18 is 5. The underlined figures are the units and the tens that we mentioned.

The new wrinkle is the same thing with a certain difference. Instead of the UT product, we now form the NT product. The N stands for "number", meaning that we use the entire number, not only the units-digit:

	N		T		
	4		3	×	6
the work	24	+	18		
the result		25			

The NT product is 25. We still have in the work 24, from 4 times 6, and 18, from 3 times 6. But now we use all of the 24, not merely the 4 part of it. We still use only the tens-digit of 18, as the letter T tells us to do. What is the NT product of 78 times 3? It is 23. Because:

	N		T			
	7		8	×	3	
the work	21	+	24			*use only underlined figures!*
the result		23				

THE DIVISION PROCESS

Two-digit divisors

First we shall go through a long example to give an overall idea of how the method works. This is for orientation only. It is not necessary that you should remember all the details at this stage. What you need to do now is observe the general way in which the process moves—the "feel" of it so to speak. It is different from what we are accustomed to in conventional long division, so we shall take a bird's-eye view of a division calculation performed by the new method. Most of the details will be postponed for a few paragraphs.

We shall divide 8,384 by 32. Using the new method, we shall eventually be able to arrive at the answer without writing any calculations at all. At this stage, of course, when we are seeing it for the first time, we prefer to write the intermediate steps. The fully worked example will be:

THE PROBLEM

	dividend	*divisor*		*answer*
	8 3 8 4 ÷	3 2	=	2 6 2
working-figures:	23 08 04			
partial dividend:	8 19 6 0			

Our first step is to take the first or left-hand figure of the dividend and make it the first of our partial dividends. Each partial dividend will lead us to one figure of the answer.

8 3 8 4 ÷ 3 2 =
↓
Partial dividend 8

The second step is to divide the partial dividend by the first figure of the divisor, the 3 of 32. The resulting figure becomes the first digit of our answer. The partial dividend may not always be exactly divisible but this is no problem as we simply ignore any remainder; thus, the first figure of our answer is 2 (8 divided by 3 is 2).

We now take this first digit of our answer and multiply it and the divisor together in a special way. This special way gives us two sets of figures. We will call these the NT figures and the U figures. (Turn back to page 143, if you need to refresh your memory on how we arrive at the NT figure.) In our present problem the NT figure for this step works out like this:

	N	T					
	3	2	×	2	=	6	
work:	06	04					
result:	06						

The U is part of an incomplete UT pair, the T being out of the picture in problems with only two-digit divisors:

		U				
	3	2	×	2	=	4
work:		04				

We are going to put NT and U figures aside for a brief moment to discuss the top row of figures in our work space. These working-figures are used only for the purpose of finding the partial dividends just beneath them:

	8	3	8	4	÷	3 2	=	2 6 2	
working-figures:		23	08	04					
		↓	↓	↓					
partial dividend:	8	19	6	0					

You will notice that each working-figure is composed of two digits, even though one may be a zero. We are going to get one digit from the partial dividend and the other from the dividend above.

The tens-digit, the 2 of 23 above, comes from subtracting the NT figure (06) we found a moment ago from the partial dividend (8).

	8	3	8	4	
working-figure:		2 ↗			
	−06 ↗				
partial dividend:	8				*partial dividend, 8, minus* NT, *06, equals tens-digit, 2, of working-figure*

To get the units figure, the 3 of 23 above, we merely bring down the next digit of the dividend:

	8	3 ↓	8	4
working-figure:		23 ↗		
	−NT ↗			
partial dividend:	8			

As we mentioned before, the working-figure exists only to lead us to a partial dividend directly beneath it. Once we find the working-figure we immediately subtract from it the U figure we found a moment ago:

	8	3	8	4	
working-figure:		23			
	−U	04 ↓			
partial dividend:		19			*working-figure, 23, minus* U *figure, 04, equals partial dividend, 19*

This new partial dividend will give us the next figure of the answer and will also lead us to the tens-digit of the next working-figure. As we mentioned at the onset of our discussion of the division process, it is important to get the "feel" of the movement of the calculation—it is the heart of the system. We start off with a partial dividend which leads to a working-figure, which leads to a partial dividend, which leads to a working-figure, and so on. Diagramatically our problem looks like this:

8 3 8 4 ÷ 3 2 = 2 6 2

working-figures:

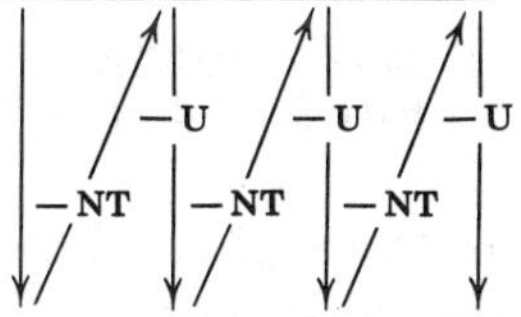

This is the heart of our system. All the rest is repetition of what we have just seen. Let us go ahead with the calculation of our example.

Our latest figure is 19. Again, we divide this partial dividend by the first figure of the divisor, that is, by 3. This gives us 19 divided by 3 equals 6. We ignore any remainders. So 6 is the next figure of the answer:

						N T		
						U		
	8	3	8	4	÷	3 2	=	2 6
work-figures:		23						
partial dividend:	8	19						*6 equals 19 divided by 32*

Then we use this 6 to multiply the 32 in two ways, first NT and then U, and make two subtractions with these results:

					N T		
					U		
8	3	8	4	÷	3 2	=	2 6
			0	*this is zero because* NT *is 19:*			
—NT							
	19						

N	T		
3	2	×	6
18	12		
19			

Carry down the next figure of the dividend:

```
8    3    8    4   ÷   3 2   =   2 6
---------------
          08
         ↗
8   19
```

Then subtract the result of the U-Style multiplication:

```
8    3    8    4   ÷   3 2   =   2 6
---------------
          08                        U
 —NT    ↗  |   —U                 3 2   ×   6
           ↓                      12
8   19     6                        U = 2
```

We get the last figure of our answer by dividing our latest partial dividend, 6, by the 3 of 32:

```
8    3    8    4   ÷   3 2   =   2 6 2
---------------
    23   08
8   19    6
```

We have now found the last figure of our answer, but we still have to determine the remainder, if any. We apply this 2 of the answer to the divisor, the 32, in the NT fashion:

```
                N  T
                3  2   ×   2
the work:      06 04
the result:     6                  our NT is 6
```

Subtracting this 6 we have:

```
8   3   8   4  ÷  3 2  =  2 6 2
               04
     —NT   ↗
         6
```

6 minus NT *6 equals 0*

We have carried down the next figure of the dividend. Now we subtract the result of the U type multiplication:

```
8   3   8   4  ÷  3 2  =  2 6 2
               04
               |
               |   —U     U
               ↓        3 2 times 2 is 04
               0   (04 − U 04 = 0)
```

This zero means that there is no remainder. The division is complete.

Is it necessary to mention that in actual work we do not draw any arrows? In practice, we would begin by writing the working-figures, but would soon find it easy to omit some of them. Eventually all the work will be done mentally: the answer being written down without any intermediate steps. At first, though, it is wise to write the working-figures as we have been doing in our example.

THE METHOD IN DETAIL

Point 1. "Doing what comes naturally" is not always recommended—sometimes it is forbidden by law. But there are many situations (and this is one of them), where the natural thing to do is also the correct thing. This is very fine, because then all you need to remember is that your first impulse is right. You divide the *first* figure of the dividend by the *first* figure of the divisor, and the result is the *first* figure of the answer. Like this:

$$\underline{8}\ 6\ 1 \quad \div \quad \underline{2}\ 1 \quad = \quad \underline{4} \qquad\qquad 4 = 8 \div 2$$

What would you do about this one?

$$1\ 6\ 1\ 2 \quad \div \quad 3\ 1 \quad = \quad ?$$

You can't divide 1 by 3. What you must do is take the first two figures of the long number, that is, 16:

$$\underline{1\ 6}\ 1\ 2 \quad \div \quad \underline{3}\ 1 \quad = \quad 5 \qquad\qquad \textit{we ignore any remainders}$$

In the same way you would have this:

$$\underline{3\ 3}\ 8\ 4 \quad \div \quad \underline{6}\ 4 \quad = \quad 5 \qquad\qquad 5 = 33 \div 6$$

Point 2. To get the other figures of the answer you continue to use the first figure of the divisor, but divide it into the partial dividends, rather than into the long number itself.

Point 3. As soon as you have found one of the figures of the answer, you use it immediately to multiply the divisor by the NT (number-tens) methods. For instance:

$$\underline{2\quad 2\quad 9\quad 4} \quad \div \quad \underline{6}\quad \underline{2} \quad = \quad \underline{3}$$

$$\downarrow$$

$$22$$

$$\begin{array}{cc} N & T \\ 6 & 2 \quad \times \quad 3 \\ \underline{18} & \underline{06} \end{array}$$

18 = the desired NT product

Multiplying the 62 by 3 in this NT style is really supposed

to be done mentally. This result, 18, is to be *subtracted from* the last figure we found, the 22:

$$2\ 2\ 9\ 4 \div 6\ 2 = 3$$

4

−NT ↗

22

$22 - \text{NT } 18 = 4$

This is the step where the rest of the long number, the dividend, comes into the picture. We bring down the next digit of the dividend, like this:

$$2\ 2\ 9\ 4 \div 6\ 2 = 3$$

↓

49

↗

22

Point 4. To make the other subtraction, you must multiply the new figure of the answer—the latest one found, the 3 here—by the units-digit of the divisor and use the units-digit of what you get:

$$2\ 2\ 9\ 4 \div 6\ \underline{2} = \underline{3}$$

49

↓ −6 *subtract 6 because 6 is the units digit of 2 × 3*

22 43

Finish the example. All we need is to repeat what we have just done:

$$2\ 2\ 9\ 4 \div 6\ 2 = 3\ \underline{7}$$

↓

43

7 = 43 ÷ 6

and then find the NT product of 62 by our new figure of the answer, 7:

2 2 9 4 ÷ 6 2 = 3 7

0

43

NT *product = 43, and 43 − 43 = 0*

and bring down the next digit of the dividend:

2 2 9 4 ÷ 6 2 = 3 7

04

43

and finally the units-digit only of 62 times the latest figure of the answer, the 7:

2 2 9 4 ÷ 6 $\underline{2}$ = 3 $\underline{7}$

04

−4

0

$2 \times 7 = 1\underline{4}$

This is the end—we have no more figures to work with. What does this last zero mean? It is the remainder. The last working figure, on the bottom line, is always the remainder.

Point 5. The last working figure on the bottom line is always the remainder. In the example above, it came out to be zero. We might say that the division "came out even." But this does not usually happen. Suppose that instead of the 2,294 of the example we had been given 2,296, and again we wished to divide it by 62. Everything would be the same except that our number is now larger by 2. We know that this extra 2 will be left over as a remainder.

Let us look at the calculation. It will all be the same except at the last step. At that step we had:

```
2   2   9   6   ÷   6   2   =   3   7
        ↓ ↗06
       43
```

What we must do now is take the units-digit of 2 (the 2 of 62) times 7, and subtract it:

```
                        U
2   2   9   6   ÷   6   2   =   3   7
           06
            ↓ −4
            2
```

So our last working figure, in the usual way of figuring, is 2. We have to stop because there is nothing left to work with, and we see that this 2 is the remainder. The extra 2 that we added on to our 2,294 (which came out even) shows up at the end as the last working figure.

Point 6. Sometimes the following will happen. We shall try to subtract the NT figure from our last working figure, as in Point 3, and we shall find that we can't do it. The number will sometimes be too large to subtract. For instance:

```
1   9   0   4   ÷   3   4   =   6          6 = 19 ÷ 3
    ↓
   19
```

Then we multiply NT style, 34 by 6:

```
  N  T
  3  4
 18+24
  20
```

```
1   9   0   4   ÷   3   4   =   6
        ?
      ↗
   −20
   /
 19
```

But this 20 can't be subtracted because we have only 19 to subtract it from.

In cases like this,

Reduce the figure of the answer by one.

Reducing our 6 by 1, the first figure of the answer is corrected to 5:

```
                    N   T
1   9   0   4   ÷   3   4   =   5
        |
        ↓
        20
      ↗
   −17
   /
 19
```

Subtract the NT *product of 34 times 5.*

From now on everything goes as usual. We subtract the units-digit of 4 times 5, which happens to be zero; then we find the next figure of the answer:

```
1   9   0   4   ÷   3   4   =   5   6
        20
        −0
        20
```

6 = 20 ÷ 3

then the NT product of 34 by this new 6 is 20—we had it before:

N T

1 9 0 4 ÷ 3 4 = 5 6

04

20

and finally we subtract the units of 4 times 6, namely 4:

1 9 0 4 ÷ 3 4 = 5 6

04

−4

0

Again we end up with a zero. There is no remainder. The answer is 56 even.

Although we have been referring to the dividend as the "long number," the dividends that have appeared in our examples have not been particularly long. We have seen four-digit numbers like the 1,904 just above. Perhaps you have been wondering whether we are restricted to such numbers.

The answer is no. The dividend can be as long as you wish, and the same method will apply. Here is a long one: 479,535 divided by 63. Spread out the figures:

N T

U

4 7 9 5 3 5 ÷ 6 3 = 7 6 1 1

39 15 13 45

−NT −U −NT −U

47 38 7 10 42

The remainder is 42.

The arrows would not be drawn in actual work. It is easy enough to imagine them there. Further: we can omit the middle line of working figures, now that we understand the idea. The work should really look like this:

							N T		
							U		
4	7	$\underline{9}$	$\underline{5}$	$\underline{3}$	$\underline{5}$	÷	6 3	=	7 6 1 1
		38	7	10	42				

Eventually, after the steps have become familiar, you will find that you will be able to do without any working figures at all. Even the single line of working figures that you see in the example just above can be omitted, if one concentrates on the work. Then nothing is written but the answer itself. The NT and the U that we showed just now are only reminders, which can be dropped just as soon as you feel that you no longer need to be reminded.

Here is a little trouble-saving device. It saves trouble in the cases that we mentioned a few paragraphs back, where we try to subtract a large NT figure from a smaller partial dividend:

> *If the second digit of the divisor is 8 or 9, don't divide by the first figure of the divisor; instead, increase the first figure of the divisor by 1 and then divide.*

For instance, if the divisor happens to be 39, we divide by 4, not by 3. The 9 of 39 has that effect. Common sense tells us why: 39 is much closer to 40 than it is to 30. Likewise, a divisor of 38 would call for dividing by 4 instead of 3. For example:

2	0	2	8	÷	3	9	=	?
	$\underline{20}$							

The first step, as we have been doing it up until now, is to say "20 divided by 3 is 6," and write the 6 as the first figure of the answer. But then we will have to correct the 6 and change it to 5, because the NT product is too big, and we can't subtract. (To be precise, the NT product of $39 \times 6 = \underline{18} + \underline{5}4 = 23$.) *Now,* however, we use the trouble-saving device and divide 20 by 4 instead of 3. Thus, the first figure of the answer is 5 immediately and will not need to be corrected:

$$
\begin{array}{ccccccccccc}
2 & 0 & 2 & 8 & \div & 3 & 9 & = & 5 & & \\
 & & & & & \downarrow & & & & & \\
 & & & & & (4) & & & & & \textit{5 = 20 ÷ 4} \\
 & \underline{20} & & & & & & & & &
\end{array}
$$

It is important to notice that we get the right answer either way, with or without the trouble-saving device. You can go still further if you wish and use this one-larger figure for the first figure of the divisor whenever the second digit is 6, 7, 8, or 9. With a divisor like 36, you would divide the working-figures by 4 to get the next figure of the answer.

If you do extend it down to 6 and 7, you will sometimes have too *small* a figure in the answer. This will need to be corrected, just as we cut down the figure of the answer when it was too large, previously. This can also happen with 8 or 9 as the second figure of the divisor, but rarely.

How can we know when the new figure of the answer is too small? Nothing about the NT product will warn us—it is small, so it can certainly be subtracted. Nothing about the U product will tell us. But here's where the "partial dividend" comes to our aid:

> *If the partial dividend is greater than the divisor, or even if it is equal to it, then the latest figure of the answer is too small.*

Suppose someone made a very careless mistake as in this division:

$$\begin{array}{cccccccc} 5 & 7 & 6 & 3 & \div & 8 & 1 & = & 6 \\ & 57 & & & & & & & \end{array}$$

This is very careless, because we know that 57 ÷ 8 is 7, not 6. But notice how this error shows itself in the partial dividend:

$$\begin{array}{ccccccccc} 5 & 7 & 6 & 3 & \div & 8 & 1 & = & 6 \\ & & 96 & & & & & & \\ & & & -6 & & & & & \\ & 57 & 90 & & & & & & \end{array}$$

NT = *48;* U = *06*

This partial dividend of 90 is obviously wrong because it is larger than the divisor. So the 6 has to be changed to 7.

If you failed to notice that 90 is larger than 81, you would have it forced on your attention at the next step. For you would say 90 divided by 8 is 11, and that would be saying "the next digit of the answer is 11." This is impossible—11 is not a digit. You would realize that 6 is too small, and you woud increase it to 7.

THREE-DIGIT DIVISORS

Suppose we wish to divide 236,831 by 674. The calculation will be almost the same as what we have been doing. It is very much as if we were dividing by 67, instead of 674. At the same time, we are going to do something about the third figure of the divisor.

You remember the diagrams we had before with upward-slanting arrows that meant "subtract NT product," and straight-down arrows that meant "subtract U product." These straight-down arrows will now have a somewhat different meaning. The new meaning brings in the new digit, the 4 of 674, as we can see by comparing these two diagrams:

TWO-DIGIT DIVISORS

N T
U
$\underline{2 \quad 3 \quad 6 \quad 8 \quad 3 \quad 1} \div 6\;7 = ?$

THREE-DIGIT DIVISORS

N T
U T
$\underline{2 \quad 3 \quad 6 \quad 8 \quad 3 \quad 1} \div 6\;7\;4 = ?$

What does this mean? Just what it seems to mean, at first glance. We form the NT product, to be subtracted on the upward arrow, by using 67 and the digit of the answer:

N T
U T
$\underline{2 \quad 3 \quad 6 \quad 8 \quad 3 \quad 1} \div 6\;7\;4 = 3$

36
↗
−20
/
23

NT = *67 × 3 =* $\underline{18} + \underline{2}1 = 20$

This is the same as before, we simply paid no attention to the 4 of 674. But now we come to the subtraction of the downward arrow—the U product. It now becomes the UT product:

N T
U T
6 7 4 × $\underline{3}$
work: $2\underline{1}$ $\underline{1}2$
result: 2

UT = *74 × 3 =* $2\underline{1} + \underline{1}2 = 2$

So in going down from the 36 that we had just now, we subtract the UT 2:

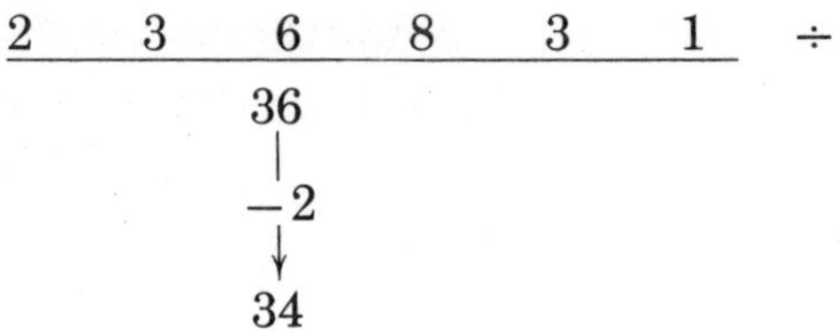

Our latest figure is 34. It is on the bottom line, so it is a partial dividend, and we shall now divide it by the 6 of 674, as usual. This gives 5 as the next digit of the answer:

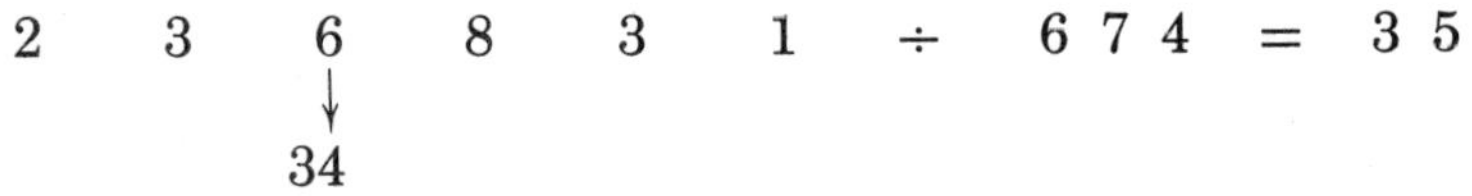

And so on. We form the NT product of 67 with this 5, and subtract from the partial dividend. Then we form the UT product of 74 with this 5 and subtract, to get the next partial dividend.

Not quite, though! The repetition is not quite exact. The next digit of the answer is not handled exactly the same as the first digit was: we use *both* of the digits of the answer already found.

The NT product is as usual: 67 times 5, is $\underline{30}$ plus $\underline{3}5$, is 33. We subtract 33 from 34, leaving 1; carry down the 8 to make 18 in the working line. But the number to be subtracted from 18 is now the sum of two parts: a UT plus a U. The diagram above shows the source of these two parts.

	U	T						U		
6	$\underline{7}$	$\underline{4}$	×	$3\underline{5}$		6	7	$\underline{4}$	×	$\underline{3}5$
	$3\underline{5}$	$\underline{2}0$						$1\underline{2}$		

result: 7, UT product, plus 2, U product, is 9

We subtract this 9 from the 18 in our division, to get a partial dividend of 9. We divide the 9 by the 6 of 674 and obtain 1. Thus, 1 is the next figure of the answer:

2	3	6	8	3	1	÷	6 7 4	=	3 5 1
		36	18						
	23	34	9						

In this particular example, no more digits need be found; 351 is the quotient.

In other problems the dividend may of course be much longer than our 236,831, and then we would have to continue with further repetitions of the process. To take care of all cases, we give this general rule:

> *Whenever the divisor has three digits (indicated by three* X*'s in the diagram below), the* UT *or "downward" subtraction is calculated as indicated:*

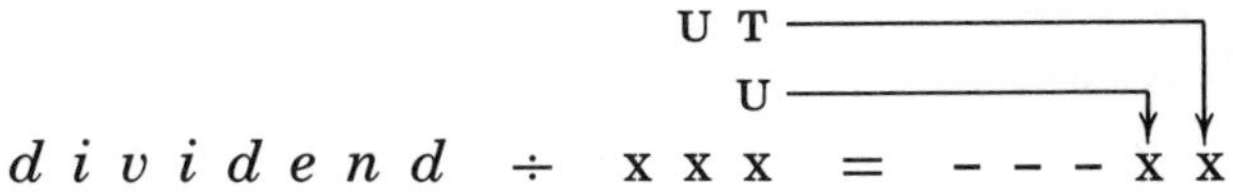

> *the two* X*'s in the answer are the two last-found digits of the incomplete answer.*

As we said, the quotient 351 was the answer in our example. However, we must still find the remainder. To know when we have the quotient in full, we do this:

> *Counting from the right-hand end of the dividend (the long number), mark off as many positions as there are figures in the divisor,* less one.

In our example the divisor is 674, which is three figures; we must mark off one less, that is, we mark off two places:

2 3 6 8 / 3 1

We observe this mark as our guide in knowing when to stop. All the figures to the left of the mark are used in finding figures of the answer, or quotient. The figures to the right of the mark are used to find the remainder. Let us find the remainder in our example:

							NT			
							UT			
							U			
2	3	6	8	/	3	1	÷	6 7 4	=	3 5 1
			18		33	261				
			9		26	257				*the remainder is 257*

We got the 33 in the working figures in the usual manner: the NT of 67 times 1, $\underline{06}$ plus $\underline{0}7$, is 6. Subtract the 6 from 9 to get 3 and bring down the 3 of the dividend. From the 33 we subtract the UT plus U products as before (UT is 74 times 1, $0\underline{7}$ plus $\underline{0}4$, equals 7; U is 4 times 5, $2\underline{0}$, equals zero) to get 33 minus 7, or 26.

The 26 is carried up without any NT subtraction, to join the 1 of the dividend as 261. The final downward step is accomplished by subtracting the product of the right-hand digit of the divisor (the 4 of 674) and the right-hand digit of the answer (the 1 of 351).

$$\begin{array}{r} 261 \\ -4 \\ \hline 257 \end{array}$$

The remainder is 257.

Think of the situation in this way: the slash through the dividend divides it into two parts: a quotient part on the left and a remainder part on the right. The slash itself is the boundary line between these regions. We cross the boundary by making an up-slanting NT arrow (subtraction), namely the one that uses the last digit of the answer. As we cross the boundary we are still acting "normally," meaning as we do in the quotient region. In fact, we complete this whole step normally, because the next downward UT + U subtraction is also done as in the quotient region. Only after that do we use the remainder region procedure, which differs from the quotient region procedure in two respects:

1. No further NT subtractions are made. The upward arrow carries up the whole partial dividend.

2. The last downward subtraction is made by using only the product of the right-hand digit of the answer (not the last two digits) and the right-hand digit of the divisor.

The description of the remainder calculation just given is a useful one—it applies, with obvious modifications, to divisors of any length. We shall come back to this point in the next section.

Here is another illustration—divide 196,307 by 512. There are three digits in 512, so we mark off two places from the right and then begin dividing:

NT
UT
U

1 9 6 3 / 0 7 ÷ 5 1 2 = 3 8 3

46 33
19 43 18

Now we cross the boundary line, the slash, with a regular NT subtraction, and complete the step with a regular downward subtraction:

NT
UT
U

1 9 6 3 / 0 7 ÷ 5 1 2 = 3 8 3

46 33 30
19 43 18 21

NT = 51 × 3 = 15
UT = 12 × 3 = 3
U = 2 × 8 = 6

Here we go into the remainder procedure—we subtract nothing on the upward arrow, instead we bring up the entire partial dividend to form part of the new working-figure. We then use only the product of the final 3 of the answer 383 and the last digit—2—of the divisor on the downward arrow (instead of using 83 as we "normally" would):

1 9 6 3 / 0 7 ÷ 5 1 $\underline{2}$ = 3 8 $\underline{3}$

46 33 30 217
19 43 18 21 211, the remainder.

Notice that to the right of the boundary line we have no "partial dividends," the numbers are all "working figures." By this we mean that none of these numbers are divided by the first figure of the divisor to give the next digit of the answer. We already have the whole answer and are now looking only for the remainder.

Here is a long one, written out as one might actually do it:

								N	T				
									U	T			
										U			
6	3	1	2	3	2 / 5	7	÷	9	8	3	=	6 4 2 1 4	
		51	32	23	62	95 897							
	63	42	21	15	48	89 895							

The answer is 64,214. The remainder is 895. (Notice that 983 is so close to 1,000, with that 8 in the second place, that we divide our partial dividends by 10, not by 9, if we wish to save ourselves the trouble of correcting some of the digits of the answer.)

Here is another one: 39,863,907 divided by 729. Because the second digit of 729 is only 2, not 8 or 9, we do not consider using 8 instead of 7 as divisor. On the other hand, at one point we shall have to cut down a figure of the answer because we run across a UT plus U figure that is too big to subtract. This is indicated by a 7 which is struck out and corrected to 6:

								N	T				
									U	T			
										U			
3	9	8	6	3	9 / 0	7	÷	7	2	9	=	5 4 ~~7~~6 8 3	
		38	66	73	39	10 07							
	39	34	50	60	22	0 0							

The answer is 54,683. The remainder is zero—it comes out even.

Examples

Here are three examples that you may like to try for yourself. They are worked out according to the "hints" that follow the third problem. Do not read the hints if you feel that you can do without them.

1. $92880 \div 432 = ?$
2. $31392 \div 654 = ?$
3. $54763 \div 489 = ?$

Hints: The last divisor, 489, has an 8 in the second position, so at each step we may divide the partial dividend by 5, instead of 4. On the other hand it would also be correct to use the 4. You may do so if you do not mind correcting the answer when necessary. In every case, remember to (1) divide each partial dividend in the respective problems, and write the result as a figure of the answer; (2) find the NT product of this figure and subtract it; and then (3) find the UT product of the latest figure of the answer plus the U product times the preceding figure of the answer and subtract that from your working figure.

ANSWERS:

1.	9	2	8	8	0	÷	4 3 2	=	2 1 5	
		12	28	08	00					
	9	6	21	0	0					*no remainder*
2.	3	1	3	9	2	÷	6 5 4	=	4 8	
			53	09	02					
		31	52	0	0					*no remainder*
3.	5	4	7	6	3	÷	4 8 9	=	1 1 1	
		14	27	66	493					
	5	6	10	49	484					*remainder is 484*

DIVISORS OF ANY LENGTH

In calculations that have divisors of four or more digits, such as in 13,671,514 divided by 4,217, we use the same basic ideas as before:

1. Subtract the NT product to find the working figure.
2. Subtract the UT product from the working figure to get the new partial dividend.
3. Divide the result (the partial dividend) by the first figure of the divisor to arrive at the next figure of the answer.

But now with a four-digit divisor we have an extra digit to take care of, such as the 7 of 4,217. We take care of it by extending the UT product, while the other two steps (**1** and **3**) remain unchanged. "Extending" the UT products means what we see in this comparison:

```
                        N T
                          U
   Two-digit divisors:  4 2

                        N T
                          U T
                            U
 Three-digit divisors:  4 2 1

                        N T
                          U T
                            U T
                              U
  Four-digit divisors:  4 2 1 7
```

Each extra digit in the divisor calls for an extra UT pair. This results in what we see above—overlapping UT pairs. A four-digit divisor will have at the peak of its development three UT pairs; a five-digit divisor, four UT pairs. We always have a single U at the end, but this is really a UT pair. The T of it is over no figure at all, so it does not contribute, and we do not bother to write that T.

What digit shall we use to mulitply each UT by? Obviously, one of the digits of the answer goes with each UT pair, but which one? Let us write an unspecified answer, or the part of it that has already been found, in the form of x's. Each x stands for some digit that we need not name here. Suppose we have already found four figures of the answer. Then the multiplication in UT style matches off the UT pairs with the digits of the answer in this manner:

Four-digit divisors:

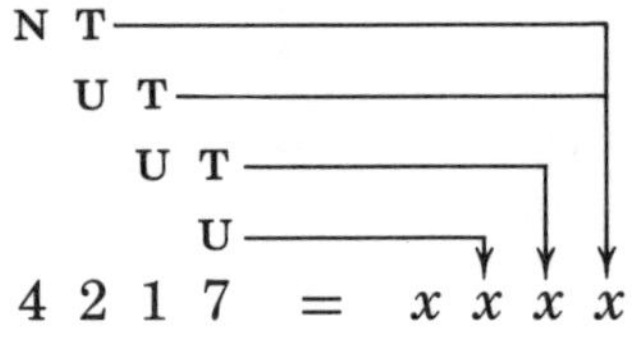

Five-digit divisors:

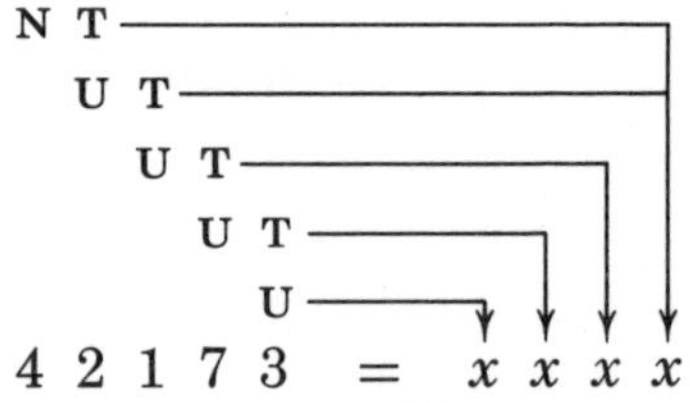

Then we *add* the results of these UT multiplications together. This is what we have been doing right along with two- and three-digit multipliers. Looking at the way the lines of the diagram fit into one another, you will notice that they form a "nested set," similar to what we had in the chapter on multiplication:

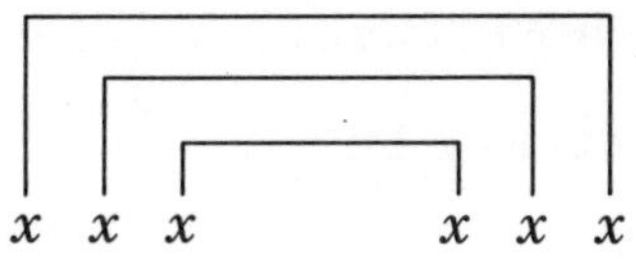

"Move inward to the middle from both ends" would be a rule that describes the motion here. The whole process can be summarized in three rules:

1. Multiply the *new* figure of the answer, at each step, by the first two digits of the divisor (like the 42 of 4,217) in NT style.

2. Multiply the same *new* figure of the answer in UT style by the second and third digits of the divisor (like the 21 of 4,217).

3. Then move in toward the middle, multiplying other UT pairs in the divisor by the "old" figures of the answer in turn (as in 4,217, we move inward twice, first to the pair 17 and then to the incomplete UT pair 7).

Let us look at the example we had above, 13,671,514 divided by 4,217. We shall apply the three rules stated just above, and see how they give the answer. First of all, we observe that the divisor, 4,217, has *four* digits, so we mark off from the right *three* places (always one less than the divisor):

```
1  3  6  7  1/5  1  4   ÷   4 2 1 7  =  3
  13
```

The last three figures of the dividend will be used for finding the remainder. The first figure of the answer is 3. It is, of course, 13 divided by the 4 of 4,217. Now we apply the three rules:

Second digit of answer:

Subtract NT product (rule 1):

```
                                     N T ─────────┐
1  3  6  7  1/5  1  4   ÷   4 2 1 7  =  3
      1
     ↗
─NT(12)                                    NT = 42 × 3
   /                                            12  06
 13                                        NT = 12
```

Subtract UT product (rule 2):

```
                                     N T ─────────┐
                                       U T ───────┤
1  3  6  7  1/5  1  4   ÷   4 2 1 7  =  3
      16
       |
  ─UT(06)                                  UT = 21 × 3
       ↓                                        06  03
  13  10                                   UT = 6
```

Divide the partial dividend 10 by 4 of 4217 and you have the next figure of the answer: 2

Third digit of answer:

Subtract NT product (rule 1):

```
                                     N T ────────────┐
                                       U T ──────────┤
                                         U T ──────┐ │
1  3  6  7  1/5  1  4   ÷   4 2 1 7  =  3 2
      16  27
         ↗
   ─NT(08)                                 NT = 42 × 2
      /                                         08  04
  13  10                                   NT = 8
```

Subtract UT products (rules **2** and **3**):

```
                                   N T
                                     U T
                                       U T
1  3  6  7  1/5  1  4   ÷   4 2 1 7  =  3 2
     16 27
         |
     —UT(09)
         ↓
   13 10 18
```

UT = 21 × 2; 17 × 3
04 02   03 21
UT = 4 plus 5 = 09

Then, of course, the next figure of the answer comes from dividing this last figure, 18, by the 4 of 4,217 (18 divided by 4 is 4).

Last digit of answer:

The fourth digit of the answer is the last digit in this example. (We know this because of the slash that divides the quotient part of the dividend from the remainder.) Subtract the NT product (rule **1**):

```
                                   N T
                                     U T
                                       U T
                                         U
1  3  6  7  1/5  1  4   ÷   4 2 1 7  =  3 2 4
     16 27 21
            ↗
     —NT(16)
          /
   13 10 18
```

NT = 42 × 4
16 08
NT = 16

Subtract UT products (rules **2** and **3**):

```
                                     N T ─────────────┐
                                       U T ───────────┤
                                         U T ───────┐ │
                                           U ─────┐ │ │
1  3  6  7  1 / 5  1  4    ÷     4 2 1 7   =    3 2 4
      16 27 21
             |                 UT = 21 × 4; 17 × 2; 7 × 3
       − UT(12)                      08 04    02 14    21
             ↓                 UT = 8 plus 3 plus 1 = 12
   13 10 18  9
```

The last figure of the answer is 2—namely, the last partial dividend, 9, divided by 4 of 4,217.

The remainder:

After we have all the digits of the quotient (the answer), we continue on to find the remainder. The general process, for a divisor of any length, is similar to what we had in the section on three-digit divisors. There are three steps:

1. We cross the "boundary" (the slash in the dividend) still calculating in the normal manner. We make our NT subtraction to find the first working number on the remainder side of the slash. The UT subtraction is also handled in the regular manner.

```
                                     N T ───────────────┐
                                       U T ─────────────┤
                                         U T ─────────┐ │
                                           U ───────┐ │ │
1  3  6  7  1 / 5  1  4    ÷     4 2 1 7   =    3 2 4 2
      16 27 21 15
              ↗|                 NT = 42 × 2
    − NT(08) / | − UT(14)              08 04
            /  ↓                 NT = 8
   13 10 18  9 1
                                 UT = 21 × 2; 17 × 4; 7 × 2
                                       04 02    04 28    14
                                 UT = 4 plus 6 plus 4 = 14
```

This was done just as before. (We will feel the effect of the remainder procedure only in the next step.) Notice how, in the divisor and quotient, the action of the NT and UT calculations move from left to right. In this step, the 3 of the quotient, 3,242, was not used. This left to right movement will continue, as we shall see below.

2. Across the "boundary," we now make use of the first of the two remainder features. From here to the end of the problem we no longer make any NT calculations. Instead we merely bring up the new partial dividend to form part of the new working figure.

U T
U T
U

1 3 6 7 1 / 5 1 4 ÷ 4 2 1 7 = 3 2 4 2
16 27 21 15 11

13 10 18 9 1

3. Finally, the last of the two remainder features: on each new downward subtraction we stop using one digit of the quotient, beginning at the left. Notice how this left-to-right movement eliminates one UT calculation at each step. Compare diagrams above and below.

U T
U

1 3 6 7 1 / 5 1 4 ÷ 4 2 1 7 = 3 2 4 2
16 27 21 15 11

−UT(11)

13 10 18 9 1 0

UT = 17 × 2; 7 × 4
02 14 28
UT = 03 plus 08 = 11

As there are no longer any NT subtractions we bring up the partial dividend figure as in step **2** above. Then, as in step **3,** we stop using the next left-hand digit in the quotient. This is the last step in our problem and, as we would expect, we end up by multiplying the units-digit (the right-hand digit) of the divisor by the units-digit of the quotient. We have continued our left to right movement and have now "run off the ends" of both divisor and quotient:

```
                                              U ──────────┐
1  3  6  7  1 / 5  1  4    ÷    4 2 1 7   =   3 2 4 2
      16 27 21 15 11 04
                      ↗ │
                     /  │ —U(04)                U = 7 × 2
                    /   ↓                             14
                                                  U = 04
   13 10 18  9  1  0  0
```

The last figure found in the calculation is the remainder. In this example, that last figure is zero. The division comes out even, so there is no remainder.

All the details we have just seen—the careful naming of the digits involved at each step—tend to make the division process seem difficult. But this is deceptive. It looks complicated only because we repeated all the small bits of work for the sake of clarity. In actual work, after the method is clearly understood, the calculation goes fast and is really quite easy.

The only real difficulty is the one that is always present—i.e., the need for being careful. Careless slips are a danger in any kind of calculation. In the division process of the Trachtenberg method, we must take care to find the correct UT pair products. Let us repeat the three rules stated on page 169 which tell us how to identify the digits in the answer by which we multiply the NT and UT pairs in the divisor.

1. Multiply the *new* figure of the answer, at each step, by the first two digits of the divisor (like the 42 of 4,217) in NT style.

2. Multiply the same *new* figure of the answer in UT style by the second and third digits of the divisor (like the 21 of 4,217).

3. Then move in toward the middle, multiplying other UT pairs in the divisor by the "old" figures of the answer (as in 4,217, we move inward twice, first to the pair 17, and then to the incomplete UT pair 7).

Perhaps the following diagrams, which show only the divisor and quotient of the problem we just finished, may give you a clearer view of the relationship of the NT and UT pairs with each new digit of the answer that is found (**1–4**) and in the determination of the remainder (**5–7**).

1.

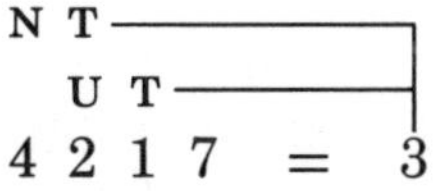

2.

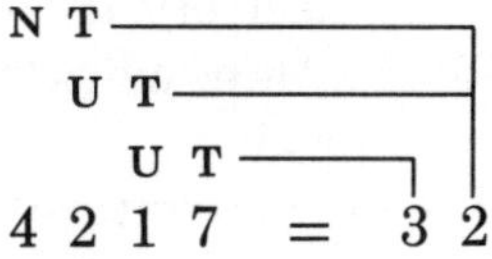

3.

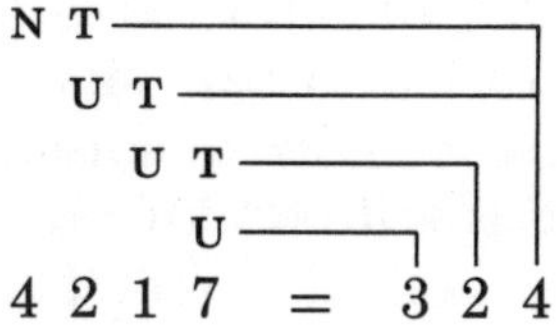

4.

```
N T
  U T
    U T
      U
4 2 1 7  =  3 2 4 2
```

5.

```
  U T
    U T
      U
4 2 1 7  =  3 2 4 2
```

6.

```
    U T
      U
4 2 1 7  =  3 2 4 2
```

7.

```
      U
4 2 1 7  =  3 2 4 2
```

Care must be taken in the handling of the numerous UT products. To reduce the danger of error, and also to save some mental effort, do this: as soon as you find each new UT product, subtract it immediately from the working figure and use what remains as a new working figure from which you subtract the next UT product.

CHECKING THE DIVISION

We have just learned two new ways to perform divisions, one of them a "simple" way and the other a "fast" way. The simple method is self-checking almost entirely, but we left one part of it for a final check. The fast method is only partly self-checking. At this point we need a systematic check on the answer and on the remainder—especially on the remainder.

Of the several possible methods of making a check, probably the most natural and most convenient is the one that follows. A possible variation of it is suggested at the end, and each individual may make up slight variations of his own, but the recommended procedure goes along these general lines:

1. Subtract the remainder from the dividend. For instance, in one example we divided 2,296 by 62, and we found an answer of 37, with a remainder of 2. Subtract this 2 from the 2,296. The result, 2,294, is a reduced dividend which would leave no remainder if divided by the 62 of our example.
2. Find the digit-sum of this reduced dividend by adding the digits across, as we have done before. The example, 2,294, gives 2 plus 2 plus 9 plus 4 equals 17. Reduce this to a single figure by adding it across: 1 plus 7 is 8. The digit-sum is 8. *Always* reduce the digit-sum to a single figure in this way.
3. Find the digit-sum of the divisor—62, in the example—and the digit-sum of the answer, the 37. Then multiply these two digit-sums together. The 62 gives 8, and the 37 gives 10, which reduces to 1. Then multiply these two together: 1 times 8 is 8. (This should be reduced to a single figure if necessary, but 8 happens to be a single figure already.)
4. Compare this product of two digits which you have just found, the 8, with the digit-sum of the reduced dividend, in paragraph **2.** That was also an 8. So 8 equals 8, and the work checks. The answer and the remainder are both correct.

One of our examples came out "even," with no remainder; 1,904 divided by 34. Here is how it would look with the checking figures written in parentheses:

(5) (7) (~~11~~) (2)
1 9 0 4 ÷ 3 4 = 5 6

Check: 7 times 2 is 14, and 1 plus 4 is 5. Compare this 5 with the digit-sum of 1,904. It is 5 also. The work checks.

What does this all amount to? It is essentially a matter of checking the inverse process, the multiplication. We saw that 1,904 divided by 34 is 56. In reverse, we could say that 34 times 56 is 1,904. This is saying the same thing, but in terms of multiplication.

Numbers:	3 4 × 5 6 = 1 9 0 4
digit-sum:	7 × 2
	$7 \times 2 = 14$, $(1 + 4) = 5$ *Check*

When there is a remainder we must get rid of it before we check the work. This is done by subtracting it from the dividend:

digit-sums	(7, see below)		(2)		(8)
numbers	6 3 1 2 3 2 5 7	÷	9 8 3	=	6 4 2 1 4
					remainder 895

63123257
−895
63122362, digit-sum = 7

$7 \longleftrightarrow 2 \times 8 = 16$
$7 \longleftrightarrow 16$

The variation that we mentioned before is optional. It helps, though, because it eliminates the subtraction at the right, above, where we subtracted the 895. This is the variation:

1. Do not subtract the remainder from the dividend, such as 895 from 63,123,257.
2. Instead, find the digit-sum of the remainder, and subtract that from the digit-sum of the dividend. (Add 9, if necessary, to make the subtraction possible.) In the example, the remainder of 895 gives 8 plus 5 is 13, or 1 plus 3 is 4. The digit-sum of the dividend is 6 plus 3 plus 1 plus 2 plus 3 plus 2 plus 5 plus 7 = 20, or 2 plus 0 is 2. Now we must subtract the 4 (remainder) from the 2 (dividend). Since the 2 is too small, we increase it by adding 9, and it becomes 11. Remember, *in digit-sums nines do not count.* They are the same as zeroes. So we subtract the 4 from 11, and we have 7.
3. As before, we obtain a number to compare with the 7. We multiply the digit-sum of the answer, 8, by the digit-sum of the divisor, 2, and we have 16. Then 1 plus 6 is 7.
4. Compare the two numbers that we have found, in paragraphs **2** and **3:** 7 equals 7. The work checks.

Practice problems

1.	$5678 \div 41$	**13.**	$81035 \div 95$
2.	$4871 \div 74$	**14.**	$63000 \div 72$
3.	$70000 \div 52$	**15.**	$4839 \div 64$
4.	$7389 \div 82$	**16.**	$2014 \div 56$
5.	$9036 \div 36$	**17.**	$5673 \div 72$
6.	$36865 \div 73$	**18.**	$5329 \div 95$
7.	$22644 \div 51$	**19.**	$4768 \div 92$
8.	$28208 \div 82$	**20.**	$5401 \div 67$
9.	$14847 \div 49$	**21.**	$2001 \div 45$
10.	$11556 \div 36$	**22.**	$7302 \div 86$
11.	$18606 \div 31$	**23.**	$9345 \div 99$
12.	$43271 \div 72$	**24.**	$85367 \div 26$

25. 479535 ÷ 63
26. 236831 ÷ 674
27. 543765 ÷ 823
28. 234876 ÷ 632
29. 204356 ÷ 913
30. 743567 ÷ 256
31. 4536754 ÷ 543
32. 27483624 ÷ 6211
33. 63123257 ÷ 9832

Answers

```
                      N T
                        U
1.  5   6   7   8   ÷   4 1   =   1 3 8
       16  37  28
    5  15  34(20)

                      N T
                        U
2.  4   8   7   1   ÷   7 4   =   6 5
           47  61
       48  43(61)

                          N T
                            U
3.  7   0   0   0   0   ÷   5 2   =   1 3 4 6
       20  30  40  10
    7  18  24  32  (8)

4.  7   3   8   9   ÷   8 2   =   9 0
           08  09
       73   0  (9)

5.  9   0   3   6   ÷   3 6   =   2 5 1
       20  03  06
    9  18   3  (0)
```

6.	3	6	8	6	5	÷	7 3	=	5 0 5
			08	36	05				
		36	3	36	(0)				
7.	2	2	6	4	4	÷	5 1	=	4 4 4
			26	24	04				
		22	22	20	(0)				
8.	2	8	2	0	8	÷	8 2	=	3 4 4
			42	40	08				
		28	36	32	(0)				
9.	1	4	8	4	7	÷	4 9	=	3 0 3
			08	14	07				
		14	1	14	(0)				
10.	1	1	5	5	6	÷	3 6	=	3 2 1
			15	05	06				
		11	7	3	(0)				
11.	1	8	6	0	6	÷	3 1	=	6 0 0
			06	00	06				
		18	0	0	(6)				
12.	4	3	2	7	1	÷	7 2	=	6 0 0
			02	07	71				
		43	0	7	(71)				
13.	8	1	0	3	5	÷	9 5	=	8 5 3
			50	33	05				
		81	50	28	(0)				
14.	6	3	0	0	0	÷	7 2	=	8 7 5
			60	40	00				
		63	54	36	(0)				

```
15.  4  8  3  9   ÷  6 4  =  7 5
           43 39
        48 35(39)

                     N T
                       U
16.  2  0  1  4   ÷  5 6  =  3 5
           41 54
        20 33(54)

17.  5  6  7  3   ÷  7 2  =  7 8
           67 63
        56 63(57)

18.  5  3  2  9   ÷  9 5  =  5 6
           62 09
        53 57 (9)

19.  4  7  6  8   ÷  9 2  =  5 1
           16 78
        47 16(76)

20.  5  4  0  1   ÷  6 7  =  8 0
           10 41
        54  4(41)

21.  2  0  0  1   ÷  4 5  =  4 4
           20 21
        20 20(21)

22.  7  3  0  2   ÷  8 6  =  8 4
           50 82
        73 42(78)

23.  9  3  4  5   ÷  9 9  =  9 4
           44 45
        93 43(39)
```

24. 8 5 3 6 7 ÷ 2 6 = 3 2 8 3
15 23 16 17
8 7 21 8 (9)

25. 4 7 9 5 3 5 ÷ 6 3 = 7 6 1 1
39 15 13 45
47 38 7 10 (42)

N T
U T
U

26. 2 3 6 8 3 1 ÷ 6 7 4 = 3 5 1
36 18 33 261
23 34 9 26 (257)

27. 5 4 3 7 6 5 ÷ 8 2 3 = 6 6 0
53 17 66 585
54 50 6 58 (585)

28. 2 3 4 8 7 6 ÷ 6 3 2 = 3 7 1
54 18 47 406
23 45 10 40 (404)

N T
U T
U

29. 2 0 4 3 5 6 ÷ 9 1 3 = 2 2 3
24 43 85 766
20 22 35 76 (757)

30. 7 4 3 5 6 7 ÷ 2 5 6 = 2 9 0 4
24 13 15 16 147
7 23 1 11 14 (143)

31. 4 5 3 6 7 5 4 ÷ 5 4 3 = 8 3 5 4
23 36 37 65 534
45 19 30 27 53 (532)

N T
U T
U T
U

32. 2 7 4 8 3 / 6 2 4 ÷ 6 2 1 1 = 4 4 2 4
34 28 43 76 622 6164
27 26 16 31 62 616 (6160)

33. 6 3 1 2 3 / 2 5 7 ÷ 9 8 3 2 = 6 4 2 0
51 32 13 32 185 1817
63 42 20 3 18 181 (1817)

CHAPTER SIX

Squares and square roots

INTRODUCTION

Here are diagrams of three ranches out west. They must be in Texas, because each one is several miles long on each side. They happen to be perfectly square in shape because these are fictitious ranches and we can make them square if we wish:

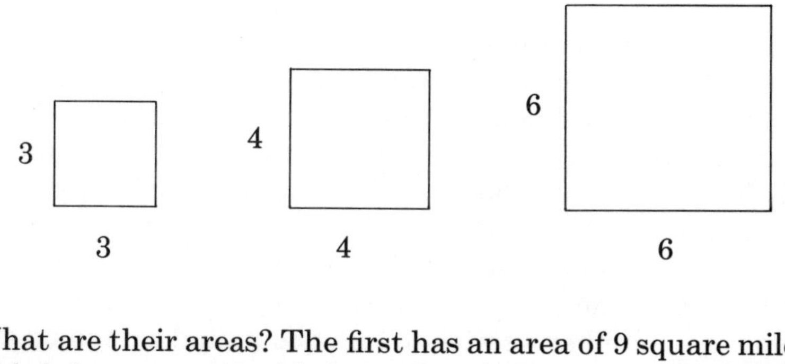

What are their areas? The first has an area of 9 square miles. This is because each side is 3 miles, and 3 times 3 is 9. In the same way the area of the second is 16 square miles, and the area of the third is 36 square miles.

Notice the arithmetic involved here. In each case it consists of multiplying a number by itself. This action is called "squaring" the number. Squaring 3 we have 9. This mathematical operation comes up as part of various problems. The simplest and most natural way in which it comes up is just what we had here, the question of finding the area of a square. For that reason it is natural for everyone to speak of it as squaring the number and to call the result of doing so the square of the number. For instance:

NUMBER	SQUARE OF THAT NUMBER
1	1
2	4
3	9
4	16
......	......
15	225
......	......
100	10,000

So we see that squaring defines a mathematical "operation." The idea of an operation is this: we operate on a number when we change it to another number. Many simple examples are familiar to everyone—for instance, doubling. We double 12, and it becomes 24. The simplest operation of all is probably that indicated by the instruction "increase by one." We operate in this way on 12, and it becomes 13, and so on. In each case we start with a particular number, in fact any number we wish to talk about, and we end up at a different number. We move from one number to another one.

Suppose we take the result of doubling, like the 24 that we mentioned, and we apply a new operation, "halving." We take half of 24 and we are back at 12, where we started. Doubling and halving are opposite operations, in that sense. We say that halving is the "inverse" operation to doubling. What is the operation inverse to "increase by one?" Obviously, "decrease by one." Apply this to the 13 of the last paragraph, and we are back at 12 again.

There is an operation inverse to squaring. It is the operation of "taking the square root." We expressed the other operations in the form of commands, "double," or "increase by one." If we express the idea of the square roots in this way, it is "answer the question: what number when multiplied by itself becomes the given number?" Examples:

NUMBER	SQUARE ROOT OF THE NUMBER
1	1
4	2
9	3
......	......
225	15
......	......
10,000	100

In this chapter we are going to consider both of these operations, squaring and taking the square root. The easier one is squaring, so we shall look at that first. It also acts as an introduction to the process of taking the square root, which is not quite as easy but has more practical value. We shall find that squaring—multiplying a number by itself—is very similar to the method of fast multiplication that we have already seen. In fact it is really a special kind of multiplication. Taking square roots is more like division.

SQUARING

Two-digit numbers

It is easy to find the square of a two-digit number, like 73. It is even fairly easy to find it by straightforward multiplication. We could say 73 times 73 and use our method of fast multiplication from a previous chapter:

$$\begin{array}{llll} 0\ 0\ 7\ 3 & \times & 7\ 3 \\ 5\ 3\ 2\ 9 \end{array}$$

But we shall now develop an even faster and easier way to get this result. Easy as it is, we may as well take it in three steps because the first two steps are interesting points in themselves. These first two steps are two special kinds of numbers, namely:

Special Type Number 1: These are the numbers which end in a 5, like 35 and 65. We can write down the square of such a number instantly:

1. The last two figures of the square are 25. This is true of any number of this special type. The square of 35 is actually 1225. In writing down the square of it first write 25 with room in front of it: 35 squared is __25.

2. To find the two figures that go in front of this 25, multiply the first digit of the given number by the next larger digit. In the case of 35, the first digit is 3, so we multiply 3 times 4. This gives 12. We put the 12 before the 25. The answer is 1225.

In the case of 65, we can think of it in this way:

$$\begin{array}{cc} 6 & 5 \\ (6\times 7) & \big| \\ \downarrow & \downarrow \end{array}$$

THE ANSWER: 4 2 2 5

Special Type Number 2: These are the numbers in which the tens-digit is 5, like 56, for instance. We write down the square of such a number immediately:

1. The last two digits of the answer are the square of the last digit of the number. With 56, we have __36, because 6 times 6 is 36.

2. The first two digits of the answer are 25 *plus* the last digit of the given number. With 56 we get 25 plus 6 is 31. This 31 goes in front of the 36 and the answer is then 3,136.

If the last digit of the given number happens to be small, as in 51, we still square it: 1 times 1 is 1. But since this has to give us the last digit of the answer, we write the 1 as 01:

```
    5     1
 (25+1)   |
    ↓     ↓
  2  6  0  1
```

The answer is 2,601. We have to have four figures in the answer, and this is how we arrive at it.

Perhaps you have noticed that we could have written these from left to right, if we had wished. There is never any number to "carry." With other numbers this will not always be true, but in these two special types of numbers there will never be anything to carry.

Now we go on to two-digit numbers in general, not restricted to any special type. We can still use two features of the special types:

1. In finding the last two digits of the answer, we shall still square the last digit of the number (like finding __25 from 35).

2. In finding the first two digits of the answer we shall still need to square the first digit of the number (like the 25 plus 1 out of 51).

We shall no longer use the other features of the special types. Instead, something new is added. Another number comes into the picture:

3. We shall now need to use the "cross-product." This is what we get when we multiply the two digits of the given number together. In squaring 34, the cross-product is 12, because 3 times 4 is 12. We shall now see how we should use this. What is the square of 32?

First step: Write 32 squared in this way: 32^2. The small 2 above the line means that we have two 32's multiplied together. (If we refer to 32 times 32 times 32, we may write it as 32^3, because there are three 32's multiplied together.) Square the right-hand digit:

$$\begin{array}{r} 3\ 2^{2} \\ \hline 4 \end{array}$$

because 2 times 2 is 4

Second step: Multiply the two digits of the number together and double: 3 times 2 is 6, doubled is 12:

$$\begin{array}{r} 3\ 2^{2} \\ \hline {}^{\bullet}2\ 4 \end{array}$$

We write the 12 as 2, and a dot to carry the 1. Squaring is very much like multiplication.

Last step: Square the left-hand figure of the number:

$$\begin{array}{r} 3\ 2^{2} \\ \hline 1\ 0\ {}^{\bullet}2\ 4 \end{array}$$

THE ANSWER: 1 0 •2 4 *3 times 3 plus the dot equals 10*

What is the square of 84?

Step 1:

$$\frac{8\ 4^2}{\dot{}6}$$

4 times 4 is 16

Step 2:

$$\frac{8\ 4^2}{{}^{6}5\ 6}$$

8 times 4 is 32, doubled, is 64 plus dot

Step 3:

$$\frac{8\ 4^2}{7\ 0\ 5\ 6}$$

8 squared is 64, plus the carried 6 is 70

Let's make a game of it—let us do the whole thing mentally. It is possible to do all the work in a purely mental fashion, without writing any calculations or even the answer. All we need is a little trick of concentration. Looking at the given number, say the 32 that we had before, we visualize three figures of two digits each:

$$\frac{3\quad 2}{09\ \ 12\ \ 04}$$

As usual we put a zero in front of a single digit like 4, to make a two-digit number out of it.

Where did these three numbers come from—09, 12, and 04? Obviously, the 9 is 3 times 3, and the 4 is 2 times 2. The 12 is twice 3 times 2. Then we "collapse" these three two-digit numbers mentally:

0(9 1) (2 0)4
add add

Add the figures in brackets: 9 plus 1 is 10, (zero and a dot); 2 plus zero is 2. We replace each bracket by the sum of the digits inside it:

0(9 1)(2 0)4
0 ·0 2 4, *that is, 1,024*

You can even do this from left to right without much trouble, after doing a few examples to get the feel of it. Going from left to right we often have to carry a figure, as we did with the 1 just above, and going from left to right we have to make a correction when we carry anything. It is easy enough to do that, because we have only these three two-digit numbers to collapse, and a little practice will enable us to hold them before us mentally at the same time. Then, when we come to a carried one like the example above, we say "9" and change it to "10" without any great effort.

It is probably easiest to calculate the cross-product and double it as our first step, and then do the squaring of the two digits. In the example above, the most convenient way is probably to look at the 32 and say "3 times 2 is 6, doubled, is 12," our middle number, and only then do we say "3 squared is 9; 2 squared is 4." With proper concentration we do not actually *say* these things, even in our minds. The proper procedure is to look at the 3 and find that a 9 is coming into our consciousness. Likewise the 2 suggests to us the figure 4. But the cross-product takes two distinct steps in the minds of most people; it does not suggest itself spontaneously until we have practiced a great deal. That is why it seems best to calculate the cross-product first. However, this is of course optional.

Try one yourself. The answer follows immediately, but you can test yourself by looking at the number once and looking away, and doing the calculation mentally before you look at the answer. The square of 43 is 1,849, which we reach in this way:

```
      4  3
 ___________
 16 24 09        24 is twice 4 times 3
 1 8  4 9
```

A little while ago we had a special method for squaring a number like 35. We said "3 times 4 is 12, and then attach 25—1225." The 4 is used because it is the next larger number to 3. Now you may wonder, would our present method with the cross-product work on the number 35? Of course it would. Do it yourself—square the number 35 mentally by using the three two-digit numbers and collapsing. The answer to squaring 35 is 1,225, any way we arrive at it, and with our present method we arrive at it like this:

```
      3  5
 ___________
 09 30 25
  12   25
```

THREE-DIGIT NUMBERS

Suppose we wish to find the square of 462. We can still make use of the operations that we learned to do on two-digit numbers:

1. We can still use the square of each of the digits separately. In squaring 32, a few pages back, you will remember that we used a 9 (from 3 times 3) and a 4 (from 2 times 2);

```
        3    2
 _______________
 09    12    04      12 is 3 times 2, doubled
   1 0 2 4
```

Now with the three-digit number 462, we shall use the square of 4 (16), the square of 6 (36), and the square of 2 (4).

2. We can also still use the cross-product of the digits, and we still double it. In squaring 32, as you see just above, we had a 12 that came from the cross-product 3 times 2, doubled. Now with the three-digit number 462, we still use cross-products, but now we have several cross-products. In fact we shall pair off the three digits in every possible way.

First step: Forget about the 4 of 462 for a moment. We have left only 62, a two-digit number. We know how to square two-digit numbers, so we go ahead and square this 62:

```
                 4  6  2
               ----------
                36 24 04
COLLAPSE IT:       3 8 4 4
```

Second step: This is new. It is not part of what we had in two-digit numbers. Form an "open cross-product" by multiplying together the first and last digits of 462, that is, the 4 and the 2, and doubling: 4 times 2 is 8, and 8 doubled is 16. Add this number directly to the two left-hand digits of our working number, like this:

```
 4   6   2
-----------
 3̸ 8̸ 4 4
 5 4 4 4          add 16: 38 plus 16 is 54
```

Notice that this is not collapsing. It is a full overlap.

Last step: Now forget about the 2 of 462 for the moment. Square 46 as a regular two-digit number, *except* that you omit the 6 squared:

```
          4  6  2
      ------------
   16 48 5 4 4 4          Don't square the 6!
   2 1  3 4 4 4           that is, 213,444
```

What does all this amount to? Looking at the bare bones of the method, so to speak, we see that it is quite natural. In squaring 462 we first worked on 62 squared. Then we ignored the 2 of 462 and worked on 46 squared.

Because 62 is the "end" of 462, squaring 62 gave us the end or right-hand part of the answer. Because 46 is the "beginning" of 462, it gave us the beginning or left-hand part of the answer. But because the 46 and the 62 overlap, we have some overlapping in the middle of the answer. To be exact:

1. The number 462 contains 6 only once. Hence, we have to use 36 (6 squared) only once: when we square 46 we don't add in another 36, we have already taken care of it in squaring 62.
2. There is one new term, which does not occur in squaring either 46 or 62. It is the "open cross-product," formed by multiplying together the first and last digits of 462. We had 4 times 2 is 8, and this 8 doubled is 16. This 16 is to be added in at the middle of the number, so we added it at the left of 62 squared.

In doing an actual problem we would not expand the work with any explanation. It would look more like this:

	3 2 5	*Write only underlined figures!*
	04 20 25	
COLLAPSE TO	0 $\underline{6}$ $\underline{2}$ $\underline{5}$	
	3 0	*3 times 5 of 325, doubled*
	3 $\underline{6}$ $\underline{2}$ $\underline{5}$	
	09 12 3 6 2 5	*32^2*
COLLAPSE:	$\underline{1}$ $\underline{0}$ $\underline{5}$ $\underline{6}$ $\underline{2}$ $\underline{5}$	

Do you remember that 25 is one of our two "special types," because it ends in 5? If we had 25 alone to square, we would say 2 times 3 is 6 (3 being the next larger number after 2), and we tag on the 25: 625. Can we use this trick here? Certainly. Just remember that we really need four digits, so 625 has to be written as 0625:

		3		2		5				
					0	6	2	5		
					3	0				*open cross-product doubled*
					3	6	2	5		
	09	12			3	6	2	5		
COLLAPSE:		1	0		5	6	2	5		

Try this one for mental practice—don't write anything but the answer. To make it easier for you this first time, we choose a symmetrical number:

2 2 2

Answer?

Don't look now—the answer is just below. After you do it yourself you can check your work against this:

	2	2	2
	04	08	04
COLLAPSE:	0 4 8 4		

Then add the open cross-product 2 times 2, doubled:

2	2	2
	~~0~~ ~~4~~ 8 4	
	1 2 8 4	

		2	2	2		
	04	08	1 2 8 4			*22²*
COLLAPSE:	4	9	2	8	4	

This study of squaring numbers will give us some insight into the method of finding the square root, which follows immediately. The method is not a repetition of anything we have had already, however. It is different from anything else, as you will see.

SQUARE ROOTS

Three-digit and four-digit numbers

When we are given a number, we know that its square root will be some smaller number with this peculiarity: when we multiply the smaller number by itself, we shall obtain the given number. Given 144, we find that its square root is 12, because 12 times 12 is 144. That is what "square root" means.

If the given number consists of three digits, like 144, or of four digits, like 1,024, the square root will have two digits. (The square root of 1,024 is 32.) That is why we are taking three- and four-digit numbers together—both kinds give us two-digit answers.

EXAMPLE ONE: Find the square root of 625. It is customary to write this in symbols as

$$\sqrt{625}$$

Read "square root of 625."

First step: Counting from the right we mark off two places:

6 / 2 5

and we work on the figure or figures to the left of this bar. In this example we begin by working on 6. The principle is general: whether the number has three digits or four, it is always true that we mark off two places from the right-hand end and use whatever lies to the left of the slash. (With 1,024 we would work on 10; we have 10/24.)

Second step: From your knowledge of the multiplication table find the largest single figure whose square is not larger than the number you found in the first step. Using 6, what is this digit? It is 2. That is because 2 times 2 is 4, but 3 times 3 is 9. We can't use 3, because 9 is larger than our 6. So 2 is the first figure of the answer:

$$\sqrt{6\ 2\ 5} = 2$$

Third step: Square the first figure of the answer and subtract it from the 6:

$$\begin{array}{l} \sqrt{6\ 2\ 5} \quad = \quad 2 \\ \ \ \underline{4} \\ \ \ 2 \end{array}$$

Fourth step: Take half of this last figure (half of the lower 2), and put a zero after it, so that we have 10. Then divide this 10 by the first figure of the answer: 10 divided by 2 is 5. This is the other figure of the answer:

$$\begin{array}{l} \sqrt{6\ 2\ 5} \quad = \quad 2\ 5 \\ \ \ \underline{4} \\ \ \ 2 \end{array}$$

Now we have two figures of the answer, 2 and 5, and we know that the answer will be a two-figure number. Are we finished? No. Because:

1. We must verify the last figure, the 5. It can happen here, as it happens in division, that a figure of the answer may be too large, or too small. We may have to go back and correct it. So the 5 that we have now is not official yet. It is tentative, not final.

2. We wish to find the remainder. In most cases the answer does not "come out even," just as usually happens in division. We can determine what the "remainder" is, the excess over the next-smaller number that would come out even.

This being a situation similar to division, you will not be surprised to find that the rest of the calculation reminds you of the division method in the previous chapter. In fact it is quite similar to that part of the calculation where you determined the remainder:

Fifth step: Imagine the answer that we have found, 25, written out in the same way that we did it in squaring:

$$\frac{2\quad 5}{04\ \ 20\ \ 25}$$ *20 is 2 times 5, doubled*

Omit the left-hand pair, the 04 in this case. We only need the 20 and 25. Collapse them:

$$\begin{array}{c} 20\ \ \underline{25} \\ 2\ 2\ 5 \end{array}$$ *zero plus 2 is 2*

So we imagine the work looking like this:

$$\sqrt{6\ \ 2\ \ 5} \;=\; \begin{array}{c} 2\quad 5 \\ (2\ 2\ 5) \end{array}$$ *imagined*

Subtract the first figure of this imagined number, the underlined 2, from our working figure (the lower 2):

$$\begin{array}{l} \sqrt{6\ \ 2\ \ 5} \\ \ \underline{4}\ \,0 \\ \ 2\nearrow \end{array} \;=\; \begin{array}{c} 2\quad 5 \\ (\underline{2}\ 2\ 5) \end{array}$$

As in division, the arrow means subtract the underlined 2

Last step: Bring down the last two figures of the given

number. We found a zero by subtracting just now. After this zero we bring down the 2 and the 5 of 625:

$$\begin{array}{lcl} \sqrt{6\quad 2\quad 5} & = & 2\quad 5 \\ \underline{4}\ 02\quad 5 & & (\underline{2}\ 2\ 5) \\ 2 & & \end{array}$$

From this 025, or whatever it is in other cases, we subtract the rest of our "imagined" number. We have already used the 2 that was underlined in the $\underline{2}25$. Now we use the remaining 25:

$$\begin{array}{lcl} \sqrt{6\quad 2\quad 5} & = & 2\quad 5 \\ \underline{4}\ 02\quad 5 & & (\underline{2}\ 2\ 5) \\ 2\quad 0\quad 0 & & \end{array}$$

the remainder is zero

The work came out even. The square root of 625 is 25.

EXAMPLE TWO: Find the square root of 645. We shall have a remainder; this one does not come out even.

First step: Mark off two places: 6/4 5

Second step: Find the first figure of the answer. It is the largest number whose square is less than 6:

$$\sqrt{6\quad 4\quad 5} \quad = \quad 2$$

Third step: Subtract the square of this number (2 times 2 is 4):

$$\begin{array}{lcl} \sqrt{6\quad 4\quad 5} & = & 2 \\ \underline{4} & & \\ 2 & & \end{array}$$

Fourth step: Take half this last figure, the lower 2, and put a zero after it:

$$\begin{array}{lcl} \sqrt{6\ \ 4\ \ 5} & = & 2 \\ \underline{4} & & \\ 2 & & \\ (10) & & \end{array}$$

Divide this 10 by the figure of the answer already found: 10 divided by 2 is 5. This is the second figure of the answer, at least tentatively:

$$\begin{array}{lcl} \sqrt{6\ \ 4\ \ 5} & = & 2\ \ \ 5 \\ \underline{4} & & \\ 2 & & \end{array}$$

Fifth step (remainder and check): Using the answer as just found, we form the second and third pairs of figures as we do in squaring:

$$\begin{array}{lcll} \sqrt{6\ \ 4\ \ 5} & = & 2\ \ \ 5 & \\ \underline{4}\ 0 & & (\cancel{0}\cancel{4}\ \ 20\ \ 25) & \\ 2 & & (\underline{2}\ \ 2\ \ 5) & \textit{collapsing} \end{array}$$

and we subtract the underlined 2. Then bring down the 45 and subtract the 25 of 225:

$$\begin{array}{lcll} \sqrt{6\ \ 4\ \ 5} & = & 2\ \ \ 5 & \\ \underline{4}\ 04\ \ 5 & & (20\ \ 25) & \\ 2\ \ \underline{2\ \ 5} & & (\underline{2}\ \ 2\ \ 5) & \\ \ \ \ \ 2\ \ 0 & & & \textit{the remainder is 20} \end{array}$$

We have a square root of 25 and a remainder of 20. This remainder is "acceptable," because it is less than the answer, 25. But this is not always the case.

EXAMPLE THREE: Here is a case where the remainder is *not* acceptable:

$\sqrt{6\quad 7\quad 6} \quad = \quad 2 \quad 5$

4 07 6 (20 25)

2 2 5 (2 2 5)

5 1 *the remainder is 51*

All the work up to this remainder was identical with the calculation in the example just above. But now we have a remainder of 51. In square root the rule is: *the remainder must not be larger than twice the answer.* Here the remainder, 51, is greater than twice the answer, 25. Evidently the 5 of our answer is too small. Maybe it should be a 6, that is, the answer may be 26 instead of 25. We try it:

$\sqrt{6\quad 7\quad 6} \quad = \quad 2 \quad 6$

4 07 6 (Ø4 24 36)

2 7 6 (2 7 6) *different now!*

0 0 *the remainder is zero*

So with 26, it comes out even: the square root of 676 is 26, exactly.

EXAMPLE FOUR: Find the square root of 2,200:

First step: 2 2/0 0

Second step: $\sqrt{2\quad 2\quad 0\quad 0} \quad = \quad 4$

because 4 times 4 is 16, but 5 times 5 is 25, too big

Third step: $\sqrt{2\quad 2\quad 0\quad 0} \quad = \quad 4$

1 6

6

Fourth step: $\sqrt{2\ 2\ 0\ 0} = 4\ \ 7$

```
 1 6
 ---
   6
  (30)                30 divided by 4 is 7
```

Fifth step (remainder and check):

```
√2 2 0 0     =   4  7
 1 6 00 0     (16 56 49)
 ----
     6 0 9      (6 0 9)
       ---
```

We can't subtract! The 7 must be too large, because we can't subtract the 9 from zero. So we cut it down to 46 and try that:

```
√2 2 0 0     =   4  6
 1 6 10 0     (16 48 36)
 -----
     6 1 6      (5 1 6)
       ---
       8 4        the remainder is 84
```

We *can* subtract 16 from 100, so 46 must be the right answer.

There was one step where we took half of the last working-figure and put a zero after it. In the example just above, the 6 under the 16 was divided by 2 and a zero added (half of 6 is 3, plus zero is 30). We used 30 by dividing it by the 4 of the answer.

Sometimes it will happen that we have an odd number to take half of, like this:

```
√3 0 2 5   =   5
 2 5
 ---
   5
```

In such cases it is probably best to use the "bigger half." With 5, use 3:

$$\begin{array}{lcr}
\sqrt{3\quad 0\quad 2\quad 5} & = & 5\quad 6 \\
\underline{2\quad 5\quad} & & (2\not{5}\;\; 60\;\; 36) \\
\quad\quad 5 & & (\underline{6}\;\; 3\;\; 6) \\
\quad (30) & &
\end{array}$$

The subtraction is blocked—we can't subtract the underlined 6 from our working-figure, 5. Therefore the 6 of our answer must be too large. We try 55 instead:

$$\begin{array}{lcr}
\sqrt{3\quad 0\quad 2\quad 5} & = & 5\quad 5 \\
\underline{2\quad 5\;}\,02\quad 5 & & (2\not{5}\;\; 50\;\; 25) \\
\quad\quad 5\;\;\underline{2\quad 5} & & (\underline{5}\;\; 2\;\; 5) \\
\quad\quad\quad 0\quad 0 & & \textit{the remainder is zero}
\end{array}$$

It comes out even: the square root of 3,025 is 55, exactly.

Five-digit and six-digit numbers

We take these together because both cases give us three-figure answers. The square root of 88,246, for instance, is 296, and the square root of 674,589 is 821. Both of these are three-digit answers. There is a remainder in each case.

This is a logical place to discuss the number of digits in the square root of a number. We can tell how many digits we shall have in the square root before we find a single figure of the answer. Roughly, we shall have half as many figures in the answer as in the given number. To be exact:

1. If the given number consists of an even number of figures (as 674,589 is six-figures, and six is an even number). then its square root will have exactly half as many figures.

2. If the given number consists of an odd number of figures (as 625 is a three-figure number, and three is odd), then we make the number of figures even by increasing it by one—three is changed to four—and we take half of that. (Half of four is two, so the square root of 625 will be a two-figure number).

We can get the same result exactly in a mechanical way, without even counting, by marking off in twos from the right. Given 674,589, before we start to find its square root we can mark it off in this way: 67/45/89. We have here three blocks of figures, so we shall have three digits in the answer. Suppose we were given 88,246. We would mark it off as 8/82/46, and we still have three blocks of figures. The fact that one block contains only the single figure 8 makes no difference. So again we expect to find three digits in the answer—and of course we do, because the answer is 296.

The advantage of marking off in two's, from the right, is that it does more than tell us how many digits our answer is going to have. It also performs the first step of our method for us. This was simply to mark off the left-hand figure or figures—one figure or two—which will give us the first figure of our answer. With 88, 246, the first step is to determine that we have 8 as our first block to work with. Then we find the digit which has as its square the greatest number which does not exceed 8. This is 2. It is not 3, because 3 times 3 is 9, and 9 is greater than 8. So the first figure of the answer is 2. Notice that counting off is necessary, because we must be sure *not* to use 88 in our first step. That would lead to 9 instead of 2: the square of 9 is 81, and that is not greater than 88. As long as we mark off in two's from the right, there will be no danger of such a mistake occurring.

All three figures of the answer, in the cases we are considering now, can be found without anything new. We need only the ideas that we have been using on the numbers three and four digits long, in the preceding section. There will be a new point in the last part of the calculation, where we are finding the remainder, and this part is necessary because it provides a check on the last digit of our answer. Once in a while we must go back and reduce the last figure by one. But notice how similar this is, at the beginning, to what we have had:

Example 1: Find the square root of 207,936. Mark it off in twos: 20/79/36. We start with 20, and we shall have three digits in the answer.

FIRST FIGURE OF ANSWER: 4 times 4 is less than 20, but 5 times 5 is greater than 20. Our first figure then is 4:

$$\begin{array}{lcl} \sqrt{2\;\;0\;\;7\;\;9\;\;3\;\;6} & = & 4 \\ \underline{\;1\;\;6\;} & & \\ \quad\;\; 4 & & \\ \quad (20) & & \end{array}$$

Half of 4 is 2, and add a zero

SECOND FIGURE OF ANSWER: This 20 divided by 4 is 5:

$$\begin{array}{lcl} \sqrt{2\;\;0\;\;7\;\;9\;\;3\;\;6} & = & 4\quad 5 \\ \underline{\;1\;\;6\;} & & (\not{1}\not{6}\;\;40\;\;25) \\ \quad\;\; 4 & & 4\;\;2\;\;5 \end{array}$$

This is the same squaring method that we had before. The 40 is 4 times 5 doubled, and the 25 is 5 squared. In squaring we have three such two-digit numbers to collapse together, but in this square-root method we use only the last two. That is because, of course, the first such number has already been used. It would be 4 squared, or 16, and we subtracted 16 in our first step, above.

LAST FIGURE OF ANSWER: Subtract the 4 of this 425 going up, and the 2 of 425 going down, like this:

```
√2  0  7  9  3  6   =   4   5
 1  6 07                4 2 5
     ↗ ↓
    4  5
      (2 or 3)
      (20 or 30)
```

Taking half of an odd number, like this 5, we don't know whether to use the "smaller half" or the "larger half," the 2 or the 3. The best thing to do on the average is *split the difference.* After adding on the zero we have either 20 or 30, and we don't know which it would be better to use. Of course we cannot really go wrong, because a wrong guess on this point would clear itself up very soon. But the natural thing to do, and the one that saves us trouble more often than any other choice, is to split the difference. Instead of either 20 or 30, we use 25. Divide this by the first digit of our answer: divide 25 by 4. This gives 6, the last figure of the answer:

```
√2  0  7  9  3  6   =   4   5   6
 1  6 07                4 2 5
    4  5
      (25)
```

We have all three figures of the answer. Notice that we have used nothing different from the techniques that we had in our shorter numbers, the ones that gave us two-digit answers. The business of splitting the difference only happened to occur in this example: it occurs as readily with the shorter numbers, or longer ones.

But from now on we have the "remainder and check" part of the calculation. In this we shall use one new idea. Even that is not entirely new, it is new only in the square root calculation. We had it before, in the squaring method. It is the use of the "open cross-product," in which we multiply together the first and last digits of the three-digit answer. With our 456, we multiply 4 times 6. As in all such cross-products, we must *double* the result: 4 times 6 doubled gives 48.

$\sqrt{2\ 0\ 7\ 9\ 3\ 6}$	=	4 5 6	
$\underline{1\ \ 6}$ 07		4 2 5	16 40 25
4 5		4 8,	*4 times 6 doubled*
		6 3 6	*see below*

The 636 is the collapsed form of 56 as a cross-product. As always in square root, we omit the first of the three two-figure numbers that we had in squaring: we omit the 5 squared. (It has already been taken care of.) Then

5 6
6$\underline{0\ 3}$6
6 3 6

is our usual cross-product form.

These cross-products should always be done mentally, after we have had a little practice. It is quite easy. We do not really need to write out 5 times 6, 30, doubled is 60; 6$\underline{0\ 3}$6 collapses to 636. It is easily done in the mind. But don't become absent-minded and forget to double!

REMAINDER AND CHECK: As soon as you have the answer, you can bring down all the rest of the given number at once, like this:

$\sqrt{2\ 0\ 7\ 9\ 3\ 6}$	=	4 5 6
$\underline{1\ \ 6}$ 07 19 3 6		4 2 5
4 5 ←		$\underline{4}$ 8
		6 3 6

We struck out the 4 and the 2 of 425, because we have already used them in subtracting from our working figure. Now we use the underscored 4 of the open cross-product, the 48, as the arrows show. As soon as we have said "5 minus 4, 1," we also strike out the 4 of 48. We have finished with it. What remains? Only this:

```
5
8
6 3 6
```

Now add the vertical column: 5 plus 8 plus 6 is 19, like this:

```
  5
  8
  6
1 9 3 6
```

We subtract this result, this 1936, from the whole remaining work-figure:

```
√2  0  7  9  3  6   =   4   5   6
          19  3  6
          19  3  6
          --------
          00  0  0          the remainder is zero
```

The problem comes out even. The square root of 207,936 is 456, exactly.

Example 2: With only a sketchy explanation, here is the work for the square root of 893,304:

```
√8  9  3  3  0  4   =   9    4
 8  1 13               (72  16)
 -------
    8                   7 3 6
  (40)
```

The 7 has just now been struck out, after we used it by subtracting it from the 8, going up. Then we subtract the 3 of 736, going down:

```
√8  9  3  3  0  4    =    9  4  5
 8  1 13                  7  3  6
 ------
    8 10
      (50)
```

The 5 of the answer is 50 divided by 9 (the 9 of 94). Now that we have the whole answer, we calculate the cross-products and bring down all the rest of the long number:

```
√8  9  3  3  0  4    =    9  4  5
 8  1 13 13  0  4         -------
 -------                  7  3  6
         10  2  5
         --------
    8 10  <-------------  9 0
                            4 2 5
                          -------
          2  7  9         1 0 2 5
          -------
```

The remainder is 279. The work checks, in the sense that we did not encounter any contradictions. There was no place where we tried to subtract a number that was too large, and the remainder of 279 is smaller than our answer, 945. Everything looks reasonable. The answer is correct.

PRACTICE EXAMPLES

Here are a few practice examples that you will find interesting to try for yourself, especially if you take them in order. The earlier ones work out in an easier way than the later ones. Answers are after the last one. When you do them, you can help yourself by using these tips:

1. When the partial dividend happens to be an odd number, "split the difference," as we said earlier. If you have a 7, for instance, and you wish to take half of it, it will give you either 3 or 4. Adding the usual zero, you would have either 30 or 40, which you would then divide by the first figure of the answer (in order to get the next figure of the answer). Which should you use, 30 or 40? Neither. Use 35.

2. When the partial dividend is zero, try 1 as the next figure of the answer, not zero. This will usually save time.

3. Whenever you divide by the first figure of the answer and find 10 as the next figure of the answer, cut it down immediately to 9. It certainly can't be 10. It may even be 8.

4. This is important, and comes up pretty often—if the remainder is larger than twice the answer, try increasing the answer and see if it is possible.

Here are the examples. Find the square root of

1. 765
2. 965
3. 200
4. 683
5. 7,888
6. 4,569
7. 46,500
8. 103,456
9. 364,728
10. 900,045

Answers: These are worked out in a practical way; that is, very much as they might be done in actual work, except that a few comments have been added. In actual work some of what we show here could be omitted, after a certain amount of practice:

1.	$\sqrt{7\ \ 6\ \ 5}$			=	2 7
	4	06	5		(28 49)
		2	9		3 2 9
	3	3	6		*remainder is* 36
	(15)				

2. $\sqrt{9\ 6\ 5} = 3\ 1$

(06 01) *see tip 2, above*

9 06 5 0 6 1

6 1

0 4 *remainder*

3. $\sqrt{2\ 0\ 0} = 1\ 4$ *we must cut down 15*

(08 16)

1 10 0 0 9 6

9 6

1 4 *remainder*

(05)

4. $\sqrt{6\ 8\ 3} = 2\ 6$ *our first answer, 25, gives a remainder of 58, larger than 25. So we try 26 instead*

(24 36)

4 08 3 2 7 6

7 6

2 7 *remainder*

5. $\sqrt{7\ 8\ 8\ 8} = 8\ 8$

(128 64)

6 4 18 8 13 4 4

4 4

1 4 14 4 *remainder is* 144

6.

```
√4  5  6  9     =    6   7
                   (84 49)
 3  6 16  9          8 8 9
       8  9
    9  8  0  remainder is 80
  (45)
```

7.

```
√4  6  5  0  0     =     2   1   5
                       Ø 4 1
 4 06  5  0  0           2 0
       2  2  5             1 2 5
 0* 2  2  7  5  remainder  2 2 2 5
                is 275
```

strike out the 0 when you subtract 0 from the 0; strike out the 4 when you subtract it from 6, second column*

8.

```
√1  0  3  4  5  6      =      3   2   1
    9 03 14  5  6             1 2 4
         10  4  1               0 6
    1  1  4  1  5  remainder is 415   0 4 1
```

9.

```
√3  6  4  7  2  8      =      6   0   3
 3  6 04 17  2  8             Ø Ø 0
          6  0  9               3 6
    0  4 11  1  9  remainder is 1,119   0 0 9
```

10. $\sqrt{9\ \ 0\ \ 0\ \ 0\ \ 4\ \ 5}$ = 9 4 8

8 1 20 30 4 5 — ~~7~~ ~~3~~ 6

17 0 4 — ~~1~~ ~~4~~ 4

9 17 13 4 1 *remainder is* 1,341 — 7 0 4

1 7 0 4

SEVEN-DIGIT AND EIGHT-DIGIT NUMBERS

These will lead to four-digit answers. We are going to set up the work in a form similar to what we had with two-digit and three-digit answers. It should be realized that this form is not really rigid. We only keep to it rigidly in presenting the method for the first time. After a person has become familiar with the method he can introduce variations to suit his own taste. Mostly these variations will be omissions—the more familiar we are with the method, the more steps we can do mentally and thus save the trouble of writing them. For instance:

1. The first figure of the answer is the digit whose square is just under the first digit, or the first two digits, of the number, as in the example:

$\sqrt{1\ \ 0\ \ 3\ \ 4\ \ 5\ \ 6}$ = 3

9

1

Variation: you can find the 3 and subtract the 9 mentally:

$\sqrt{1\ \ 0\ \ 3\ \ 4\ \ 5\ \ 6}$ = 3

1

2. In finding the other figures of the answer, we have used the cross-products and "collapsed" the two-figure numbers:

```
√1   0   3   4   5   6   =   3   2
     1                     (12  04)
```

In the examples that we have been considering, we have usually shown these two-figure numbers, like 12 04, and then collapsed them. These need not be written, in actual work. Looking at 32, we can arrive at 124 mentally, after only a little practice. But don't forget to *double* the cross-product!

3. After all the figures of the answer have been found, we go on to the remainder part of the calculation. The examples above showed this as done in one subtraction. The figures in the column of the table at the right were added up, and after all columns had been added we subtracted the total in one step, from all the unused figures of the given number:

```
√1   0   3    4   5   6   =   3   2   1
             14   5   6       1̸ 2 4
             10   4   1         Ø 6
REMAINDER:    4   1   5           0 4 1
                                 10 4 1
```

Variation: we could subtract in sequence, one column at a time, first the 10, then the 4, then the 1:

```
√1   0   3    4    5    6    =   3   2   1
             14   45  416        1̸ 2 4
      (−10) |  (−4) |     | (−1)
            ↓       ↓     ↓        Ø 6
             4    41  415
                                     0 4 1
                                    10 4 1
```

This is an advantage after we have become so familiar with the method that we do not write the little table under the 321. This can be done mentally. In that case we do not calculate the little table all at once, we calculate one column at a time. As soon as we have calculated a column we use it, by subtracting it from the working-figure, so that we can immediately forget that column. But this comes naturally only after a good deal of practice.

For seven-digit and eight-digit numbers, we use everything we have already had in shorter numbers, and we add on something at the end. To be specific:

1. We find the first three of the four digits of the answer exactly as we found three-digit answers before. For instance, the square root of 10323369 is 3,213, as we shall soon see. The first three digits of the answer are 321, so the work is exactly the same as in the shorter example just above:

$\sqrt{1\ 0\ 3\ 2\ 3\ 3\ 6\ 9} = 3\ 2\ 1$

1 03   1 2 4

1   0 6

0 4 1

But now we do not go ahead with the remainder calculation. We still have another figure of the answer to find.

2. We find the fourth figure of the answer like this: we use the next column of the little table, 4, 6, 0, exactly as we always do in finding figures of the answer. That is, we find the total of the column (4 plus 6 plus 0 is 10), and we subtract the tens-digit of this total on an upward arrow:

$\sqrt{1\ 0\ 3\ 2\ 3\ 3\ 6\ 9} = 3\ 2\ 1$

1 03 02   1 2 4

↗ (−1)

1   0 6

0 4 1

then we subtract the units-digit of the total (zero) on a downward arrow:

```
√1 0 3 2 3 3 6 9  =  3 2 1
   1 03 02           1 2 4
         2             0 6
                         0 4 1
```

We have struck out the 4, the 6, and the zero, because we have used them just now—we subtracted both digits of 10.

We take half of this last 2 with a zero after it (half of 2 is 1 plus zero is 10) and we divide by the first digit of our answer, the 3 (10 divided by 3 is 3). The resulting 3 is the last figure of the answer:

```
√1 0 3 2 3 3 6 9  =  3 2 1 3
   1 03 02           1 2 4
      1  2             0 6
        (10)             0 4 1
```

3. All the unused figures of 10,323,369 go into the remainder. Here is where we need to extend the method. The little table under the answer needs more columns now. Let us postpone this for just a moment while we look at the same steps figured on a different number. Find the four figures of the answer in the square root of 40,094,224:

```
√4 0 0 9 4 2 2 4  =  6 3 3 2
  3  6 10 19         3 6 9
     4  4  3           3 6
  (20)(20)(15)           1 8 9
```

The first three figures of the answer, 633, were found just as they would be in an example with a three-digit answer. The last figure, the 2, is found by subtracting the column 9,

6, 1, which totals 16, from the 19, and taking "half" of the resulting 3. When we add the zero we split the difference and use 15. We divide it by the 6 of 633, and we have 2, the last figure of the answer.

Remainder and check

1. Bring up the last working-figure and "tag on" after it all the unused figures of the long number:

FIRST EXAMPLE

√1	0	3	2	3	3	6	9	=	3 2 1 3
	9	03	02	↗ 2 3 3 6 9					
	1	1	2						

SECOND EXAMPLE

√4	0	0	9	4	2	2	4	=	6 3 3 2
3	6	10	19	↗ 3 4 2 2 4					
	4	4	3						

From the number that we have just found, 23369, or 34224, or whatever it may be, we shall subtract the number which we shall find in the next paragraphs. The result will be the remainder.

2. We find the number that must be subtracted by extending the little table under the answer. We have already, in the first example:

3	2	1	3
~~1~~ 2 ~~4~~			
	~~0~~ ~~6~~		
		~~0~~ 4 1	

This table was made without using the last figure of the answer, the 3. What we add on now is simply the contributions from this last figure: we form cross-products of this 3 with the other figures of the answer, in all possible ways, and we square the 3:

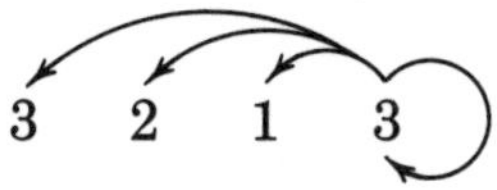

Multiplying, we have 9, 6, 3, and 9, in that order. But for our table in the square-root method all cross-products must be doubled, so we have:

```
1 8
  1 2
    0 6
      0 9
```

They go from left to right one step at a time, just as we go across 3213 from left to right one step at a time when we form them.

Now how shall we fit these new numbers into our little table? We write the very first of the digits in this four-number table, the 1 of the 18, in the last column that was struck out. But remember, any single-digit numbers have to be written with a zero in front, just as the 6 in the table is shown as 06. If our first number had been an 8 instead of an 18, we would have written it as 08. This is important—it keeps numbers in their correct columns. So in the first example we have:

```
        3   2   1   3
        -------------
        1 2 4
          0 6
            0 4 1
            1 8
              1 2
                0 6
                  0 9
        -------------
TOTAL:      2 3 3 6 9
```

Actually, the rule is this—the tens-digit of the 18, or whatever the number may be in other examples, goes in the last struck-out column. Normally this works out in a natural way, as it does with 18, because normally we do have a two-digit number. But sometimes we have a single digit, like 6. This takes care of itself if we remember to write 6 always as 06. Sometimes also we may have a three-digit number: 9 times 7 is 63, doubled, gives 126. In those cases, we still put the tens-digit in the last struck-out column. With 126, the 2 would go into that column.

In the other example that we were looking at we have:

```
6   3   3   2
-------------
3 6 9
  3 6
    1 8 9
    2 4
      1 2
        1 2
          0 4
-------------
    3 4 2 2 4
```

Subtracting this number we get zero. Both of our examples

came out even, no remainders. This does not usually happen in general when we are given a number and asked to find its square root:

Answer, 3,160

$\sqrt{9\ 9\ 8\ 7\ 6\ 3\ 4}$	=	$3\ \ 1\ \ \not{5}\ \ 6\ \ \not{1}\ \ 0$
9 09 08 5 6 0 0		$\not{0}\ \not{6}\ \not{1}$
0 3 0 2 0 3 4		$\not{3}\ \not{6}$
		$\not{1}\ 5\ 6$
		0 0
Remainder is 2,034.		0 0
		0 0
		0 0
		5 6 0 0

We took the second figure of the answer as 1, not zero, though zero divided by 3 is zero. This is a guess, based on the large digits coming next, 9 and 8. In any case, we *can't go wrong*. For suppose we had used the incorrect zero. We would have had:

$\sqrt{9\ 9\ 8\ 7\ 6\ 3\ 4}$	=	3 0
9 09		
0 9		
(45)		

What is the next figure of the answer? Divide this 45 by the 3 of the answer, and you have 15. But 15 can't be the next digit of the answer, because 15 is not a digit. It is two digits. So we know we must increase the zero that we tried, and then we have 31. In the same way the 5 of 315 was too small. We increased it to 316.

Find the square root of 8,724,321. Marking it off, we have 8.72.43.21, so we shall have a four-digit answer, and we begin with the 8, not with 87:

```
√8  7  2  4  3  2  1   =   2    9
 4 07                      4 4 1
   ↗ ↓
  4  3
(20) (15)
```

The 20 divided by 2 gave us 10 as the second digit of the answer, which is impossible. We have cut down the 10 to 9. Then we have 29 as part of the answer when we come to the 15, which was what we get when we "split the difference." Notice this point: instead of dividing the 15 by the 2 of 29, we divide the 15 by 3. This is a matter of common-sense. Any number beginning 29 is almost as large as one beginning 30, and it is far away from one beginning 20. Hence we can say that this particular 2 is almost a 3, because of the 9 after it. (Of course we would arrive at the right answer either way, it is a question of possibly wasting a little time.) So 5 is the next figure of the answer. Continuing to the end:

```
√8  7  2  4  3  2  1    =   2   9   5   3
 4 07 12 24  3  2  1        4 4 1
         20  2  0  9          2 0
 4  3  2  4  1  1  2            9 2 5
      (10)                      1 2
                                  5 4
                                    3 0
                                      0 9
                               ----------
                                2 0 2 0 9
```

remainder is 4,112

LONGER NUMBERS

Numbers still longer than those we have mentioned are handled by the same principles. The method that we have just seen for four-digit answers will give us the first four digits of longer answers. The fifth figure comes from an application of the same method. If there are only five digits in the answer we then get the remainder by using the little table under the answer, but now the table contains extra entries. They are the cross-products of the fifth figure of the answer by all the other figures in turn, ending with the square of the fifth figure. As usual, all the cross-products are doubled, but the square of the fifth digit is not doubled.

The square root of 872,079,961 is marked off as 8.72.07.-99.61, so there will be five digits in the answer, and we begin with 8:

```
√8  7  2  0  7  9  9  6  1   =   2   9   5   3
 4 07 12 10                      4 4 1
 4  3  2  1                        2 0
         (05)                        9 2 5
                  (subtract up)      1 2
                                       5 4
                                         3 0
(subtract these going down)                0 9
```

"Splitting the difference" on the 1 gives us the 05 shown as our last working figure. Divide this by 3, from the 29, and we have 1 or 2, we don't know which. Actually 1 proves to be correct, so we shall avoid rewriting by using it immediately:

```
√8  7  2  0  7  9  9  6  1   =   2   9   5   3   1
 4 07 12 10 17  9  9  6  1       4 4 1
                17  9  9  6  1     2 0
 4  3  2  1  no remainder            9 2 5
                                     1 2
                                       5 4
                                         3 0
                                           0 9
                                       0 4
                                         1 8
                                           1 0
                                             0 6
                                               0 1
                                 ------------------
                                       1 7 9 9 6 1
```

In accordance with our principles, the new slanting line of figures, 04, 18, and so on, comes from multiplying the last figure of the answer across all the other figures of the answer and doubling, and then squaring the last figure itself. Also in accordance with our principles, the tens-digit of the first number in the new slanting line, the tens-digit of the 04, is in the last struck-out column.

In practice we need not have such a spread-out array. In fact, we would not. But it is best for each individual to choose his own way of handling it. Either he can collapse some of these two-digit numbers together, which is advantageous if he writes the result, or he can omit writing the two-digit figures entirely, and instead subtract each column sequentially.

CHECKING

In squaring numbers, and in finding square roots, we can use methods of checking very similar to what we used in multiplication and division. In fact, squaring is a particular kind of multiplication, in which a number is multiplied by itself, so we can use precisely the multiplication check. This consists of finding the digit-sum of the numbers multiplied together and the digit-sum of the result, and seeing whether or not they agree. This applies directly to the squaring of numbers. Take the example 32^2 equals 1,024. We are not likely to have made an error in such an easy one, but it serves as an illustration.

$3 + 2 = 5$, the digit-sum of 32
$1 + 0 + 2 + 4 = 7$, the digit-sum of 1,024

If squaring 32 leads to 1,024, then squaring the digit-sum of 32 should lead to the digit-sum of 1,024. Does it? The digit-sum of 32 is 5. Square it and you have 25, which reduces to 7. Remember, *all digit-sums must be reduced to a single figure.* So squaring the digit-sum of 5 gives us 7. Compare this with the digit-sum of 1,024—it is also 7. They agree, and the work checks.

Checking square roots. In this we do the same sort of thing that we did in checking division. We check the inverse process instead, which is fully equivalent to checking the process itself. For example, one of the square-roots that we worked out was the square root of 207,936. We found it to be 456 exactly, with no remainder. Let us check it:

$$\sqrt{2\ 0\ 7\ 9\ 3\ 6} = 4\ 5\ 6$$

Digit-sums: 0 6

Square the 6, and you have 36. But 3 plus 6 is 9, which is zero, in digit-sums. So the two agree. The work checks.

The reasoning, of course, is this: to say that the square root of 207,936 is 456 is the same thing as saying that 456 squared is 207,936. One is true if the other is. So we check the squaring of 456, rather than try to take the square root of the longer number. You couldn't tell in advance whether you should use the square root of zero, or of 18, or of 27, or of 36, etc. They all equal one another, in digit-sums. This way, we square the digit-sum of the root, and we have a reliable check.

What do we do when there is a remainder? The same thing that we did in division. We trim off the remainder, either actually or in the digit-sums. For instance, we worked this example earlier:

	$\sqrt{4\ 6\ 5\ 0\ 0}$	$=$	2 1 5
	remainder 275		
Digit-sums:	6		8
	remainder 5		
Remove remainder:	1		

To check, we *square* the 8, which gives 64, or 1, in digit-sums. The work is correct.

In the calculation of the square root, we had partial checks as the work proceeded. This final check on the whole answer and remainder, however, is still highly desirable.

CHAPTER SEVEN

Algebraic description of the method

Hardly a person is now alive, probably, who has not been challenged at one time or another to answer a question something like this:

A carpenter found one day that he had a long board, too long to use. So he took a saw and cut it into three pieces. The first piece was 3 feet long. The length of the second was equal to that of the first plus one-fourth the length of the third. The third piece was as long as the other two together. How long was the original board, and how long was each piece?

If you happen to be a puzzle fan you will certainly recognize this—it is one of the standard types of puzzles. The answer to this particular one is 16 feet for the original board, and the three pieces are 3, 5, and 8 feet respectively, as we shall see later.

Any way that you may have arrived at the result is fair, and if you did not work it at all, that is fair too. We are not interested in the puzzle for its own sake. We mention it only because it is a beautiful illustration of one point of view of algebra. This particular problem, and all of its type, can be done best by algebra. The point of view of algebra that it implies is what we may call "x is the unknown." A certain number is being sought, and we shall not know its identity until the problem is solved. So in the meantime, we call it by a letter, x or any other letter, as a sort of alias.

This, of course, is not all there is to algebra. For one thing, in the field of pure mathematics there is a very extensive and varied structure of algebraic theory, which does not try to solve problems of this kind. For another, there are applications of algebra which have more practical value and are different from the puzzle-solving kind of algebra. One of them we shall use now, and it will be useful all through this chapter. Different from the "x is the unknown" idea is the class-description point of view. It is a way of talking about a whole set of numbers all at one time, without picking out any individual number.

Consider this situation: *A certain group of men belong to a bowling team, and their wives belong to a women's bowling team. Mr. A is 28 years old and his wife is 26; Mr. B is 25 years old and his wife 23; Mr. C is 29 and his wife 27; Mr. D is 23 and his wife 21; and Mr. E is 24 and his wife 22.*

How can we summarize these figures? We can do it by noticing that the women's bowling team is two years younger than the men's team. In fact, each husband is two years older than his wife. Let us write the letter h to represent the age of any one of the husbands and w for his wife's age—choosing the first letter of each word to remind us which is which—and we have:

$$h = w + 2$$

This is the same as saying that each wife is two years younger than her husband:

$$w = h - 2$$

The same fact can also be written in a notation which uses subscripts. We write a for "age" and we put a small $_h$ or $_w$ a little below the line to indicate "husband" or "wife," and then the relation is written as

$$a_h = a_w + 2$$

To see how this can be useful, let us use the letter s to stand for "score," meaning an individual's average score for one game, and we might have this relation:

$$s_h = s_w + 25$$

This would be the case if each husband scored 25 points higher than his wife.

The point is that each of these equations describes a situation which is true for all the numbers of the set, that is, for all the members of the team. When we write $h = w + 2$, we are referring in one equation to the fact that $28 = 26 + 2$, for Mr. and Mrs. A; also to the fact that $25 = 23 + 2$, for Mr. and Mrs. B, and so on. It is a general statement which includes all the special ones.

In this very simple example we could have used words instead of the algebraic symbols. "Each husband is two years older than his wife," is easy to say and easy to grasp. But in more complicated situations we are able to handle the relations involved easily in symbols, whereas the statement in words would be formidably long and complicated. That is the kind of situation we shall encounter now, in describing the essential parts of Trachtenberg system in the language of algebra. In this chapter we shall

(1) begin by looking at part of Chapter One from a new point of view, to illustrate what we are trying to do, and then

(2) give a short review of the basic methods of algebra, which of course is not a part of the Trachtenberg system itself. This review is intended for the convenience of those who would like to refresh their memories; others may prefer to skip over it. Then

(3) we shall apply the basic ideas of algebra to the procedures already presented as the Trachtenberg method.

NUMBERS IN GENERAL

In the previous chapters we have been working with numbers, combining them in various kinds of calculations. In every case we had a particular number or pair of numbers before us, such as 4,776 multiplied by 63. We did not make any use of letters to represent numbers, as the a_h and a_w represent the ages of husband and wife.

Now we are going to look at the Trachtenberg method, or at least at the most important parts, with the aid of letters representing numbers. This makes it possible to talk about all numbers at once, and we can make statements about the method that will always be true, regardless to what particular number we may apply our rules.

This is optional. It is not required, for any practical application, that anyone should read this chapter. On the other hand many persons find this kind of discussion interesting, and for their sake we include it. Also, there are two real benefits that can be obtained from this:

1. The algebraic formulation proves that the rules we have been using are correct. More than a few persons have a tendency toward skepticism about new ideas. In fact some of this skepticism can linger on even after a few examples have been tried and the method has given correct answers. The general method, using algebra, proves that the rules will always lead to correct answers. This will remove any doubt that may remain in one's mind, and also enable us to convince any second person who may challenge this method.

2. The algebraic formulation gives us insight into the principles that are at work. When we work a particular example, like 4,776 times 63, our attention must be focused on the figures with which we are working. The manner in which the pieces of the picture come together to give the whole result is buried under the details of the calculation. But when we use the algebraic method, representing numbers by letters, it is quite different. We do not carry out any actual calculation. We do not say "3 times 6 is 18," and so on. Our minds are free to look at the manner in which the different parts of the numbers combine to give the answer. This leads to a better understanding and a firmer grasp of the method.

In order to see how the Trachtenberg method operates, we shall need to "spread the numbers out" so that we can see what the individual figures of the numbers are doing. Suppose the number is 357. This means three hundreds plus five tens plus seven ones:

$$3\ 5\ 7 = 3 \times 100 + 5 \times 10 + 7$$

The number 704 can be written similarly, in the form

$$7\ 0\ 4 = 7 \times 100 + 0 \times 10 + 4$$

In terms of money, this means we have 7 hundred-dollar bills, *no* tens, and 4 ones.

Any number at all in the hundreds range–from a hundred to a thousand–can be written in this form:

$$a \times 100 + b \times 10 + c$$

Each of the letters, *a*, *b*, and *c*, stands for a single figure. The single figure can be zero, or 9, or any whole number in between. So the letter *a* stands for a figure from zero to 9, and *b* and *c* stand for figures of the same kind, either the same or different:

$$a = 7$$
$$b = 7$$
$$c = 7$$

This leads to the number 777:

$$7\ 7\ 7 = (7 \times 100) + (7 \times 10) + 7$$

Longer numbers, say a six figure number for example, can be written in the same way, as:

$$(a \times 100{,}000) + (b \times 10{,}000) + (c \times 1{,}000) + (d \times 100) + (e \times 10) + f$$

In all these expressions we have used the × to indicate multiplication. We read "$a \times 100$" as "*a* times one hundred." However, it is more convenient to omit the times sign. No × need be written. The more customary form is not "$a \times 100$" but rather "100*a*," which is read "one hundred *a*." The fact that the 100 and the number *a* are multiplied together is understood, simply from the fact that they are written side by side. So the six-figure number could be written as

$$100{,}000a + 10{,}000b + 1{,}000c + 100d + 10e + f$$

In the ordinary way of writing, which we use when we have figures instead of letters, this number would be *abc,def*. We need the spread-out form for the later calculations.

Further, notice that we may add a zero in front of the number if we wish, because this leaves the number unchanged. We may wish to do this in order to put an extra zero in front of a number, as we did in the chapter on multiplication. Then the number 357, for instance would be:

$$3\ 5\ 7 = (0 \times 1{,}000) + (3 \times 100) + (5 \times 10) + 7$$

Remember, *any number multiplied by zero is equal to zero.* Consequently, the $(0 \times 1{,}000)$ of the equation just above is equal to zero, and we may add it or not, as we please. We choose now to add it because we are going to describe multiplication, and in that method we place an extra zero in front of the given number. Any number at all of three digits is represented by:

$$(0 \times 1{,}000) + (a \times 100) + (b \times 10) + c$$

or else if we prefer by:

$$(1{,}000 \times 0) + 100a + 10b + c$$

THE RULE FOR ELEVEN

Now let us use this to examine the "rule for eleven" in multiplication. As you remember, the rule is simply "add the neighbor," the neighbor being the figure immediately to the right of the one you are considering at the time. It is understood that we must write a zero in front of the given number, and apply the rule to this zero also. The last figure of the given number, the digit on the extreme right, has no neighbor at all, of course, so there is nothing to add to that figure. To see how it works, let us take a four-figure number,

$$\begin{aligned} N &= (0 \times 10{,}000) + (a \times 1{,}000) + (b \times 100) + (c \times 10) + d \\ &= (10{,}000 \times 0) + 1{,}000a + 100b + 10c + d \end{aligned}$$

This represents *all* the four-figure numbers that there are, and choosing a particular four-figure number means giving particular values to *a, b, c,* and *d.* We shall not give them any particular values, because now we wish to talk about all four-figure numbers at once. What we are going to do is multiply this general number by 11 (11 is 10 plus 1):

$$1\ 1 = 10 + 1$$

So when we multiply any number by 11 we are, in effect, multiplying it by 10, and multiplying it by 1, and adding the two results:

$$\begin{aligned} 3\ 5 \times 1\ 1 &= 3\ 5 \times (10 + 1) \\ &= (3\ 5 \times 10) + (3\ 5 \times 1) \end{aligned}$$

But multiplying a number by 10 simply adds on a zero at the right-hand end of the number (35 × 10 is 350). So:

$$3\ 5 \times 1\ 1 = 3\ 5\ 0 + 3\ 5 = 3\ 8\ 5$$

"To multiply by ten, add a zero at the right" explains itself when we use the spread-out form of the numbers, and add a zero (which does not change its value!):

$$3\ 5 = (3 \times 10) + (5 \times 1) + 0$$

Times 10:

$$\begin{aligned} 3\ 5 \times 1\ 0 &= (3 \times 10) \times 10 + (5 \times 1) \times 10 + 0 \times 10 \\ &= (3 \times 100) + (5 \times 10) + 0 \\ &= 3\ 5\ 0 \end{aligned}$$

Now we do the same things to the general four-digit number—not a particular number, but any one:

$$\begin{aligned} 10 \times \mathrm{N} &= 10 \times (10{,}000 \times 0 + 1{,}000a + 100b + 10c + d) \\ &= 100{,}000 \times 0 + 10{,}000a + 1{,}000b + 100c + 10d + 0 \end{aligned}$$

Multiplying by 10 has added another zero to each of the

factors 10, 100, 1,000, and so on, which moves every figure over to the left, and leaves a zero at the right-hand end.

Now we multiply the general number by 1. *Multiplying any number by one leaves it unchanged.* So we have:

$$1 \times N = 10{,}000 \times 0 + 1{,}000a + 100b + 10c + d$$

Finally we add this expression for $1 \times N$ to the previous expression for $10 \times N$, and we have $11 \times N$:

$$\begin{aligned} 11 \times N = 100{,}000 \times 0 + 10{,}000a \quad &+ 1{,}000b + 100c + 10d + 0 \\ + 10{,}000 \times 0 + 1{,}000a &+ 100b + 10c + d \end{aligned}$$

Add these numbers in pairs, by adding each term to the one immediately beneath it, like this:

$$\begin{aligned} 11 \times N = 0 \times 100{,}000 + (a + 0) \times 10{,}000 + (b + a) \times 1{,}000 \\ + (c + b) \times 100 + (d + c) \times 10 + d + 0 \end{aligned}$$

Now it is always true, in ordinary arithmetic, that $a + b = b + a$, for any numbers a and b. For instance, $3 + 5 = 5 + 3$. Each pair equals 8. So we can reverse the order of the pairs of letters that are added together. The equation for $11 \times N$ becomes:

$$\begin{aligned} 11 \times N = 0 \times 100{,}000 + (0 + a) \times 10{,}000 + (a + b) \times 1{,}000 \\ + (b + c) \times 100 + (c + d) \times 10 + d + 0 \end{aligned}$$

This is the "rule for eleven." To multiply by 11, we take each figure of the given number in turn and we add to it its "neighbor." The neighbor of a is b, because the given number in ordinary writing is a,bcd. The neighbor of b is c. We have added the neighbor, and the equation for $11 \times N$ is the same as the rule for eleven that we have been using. This proves that the rule is correct.

Why did we put the zero in front? To take care of "carried" figures when they happen to occur. Notice that we must carry figures in the ordinary way. Suppose that b equals 7 and c equals 8. Then one part of our answer is the term $(b + c) \times 100$ and this becomes $(7 + 8) \times 100$, which is 1,000 plus 500 (or 1,500). So this term contributes not only to the "hundreds" position of the answer, but also to the "thousands" position. It contributes a 1 to the thousands. This is a "carried" 1 in ordinary language. The thousands term of the answer is $(a + b) \times 1{,}000$ in the equation above. But we have carried the 1 from the hundreds term, and it becomes $(a + b + 1) \times 1{,}000$ when we have $b = 7$ and $c = 8$. This shows that we must carry over the 1 of 15, or whatever it may be, to the next higher position.

Whenever the given number happens to be in the 9,000 range we expect to have a carried 1 at the last step. This explains the need for a zero in front. Suppose that $a = 9$ and $b = 8$, so that the given number is in the 9,800 range. We multiply it by 11. What is in the thousands place? It is $a + b = 9 + 8 = 17$, plus possibly a carried number if c is large. It is at least 17. We must carry at least a 1, possibly a 2. Then what is in the ten-thousands place? It is either $0 + a + 1$ or else $0 + a + 2$, and since $a = 9$ in the example, it is either 10 or 11. In either case we must carry over a 1 into the hundred-thousands place. This place becomes $(0 + 1) \times 100{,}000$. We see that the zero in front of the given number has provided us with a place to put the carried 1. That is all it does, but that is enough. If we should ever forget the carried figure, the answer would be disastrously wrong.

This takes care of all four-digit numbers. What shall we say about five-digit numbers, and the rest? There are two ways that we can handle them, both satisfactory:

1. Observing that we have not made any use of the fact that our number was four digits long, we may simply say "the same argument obviously holds true for numbers of any length." A five-digit number, for instance, would have one more letter; it would look like *ab,cde*. But the way we multiply by 10, and the addition to this of the number itself, and the grouping of the letters by pairs as $(a + b)$ and so on, would all go through in the same way. The same argument actually does hold good.

2. There is a neat way of writing numbers of any length, and using this notation takes care of everything. We shall see this a little later. It is not really necessary at this point, and it is more convenient to postpone it.

ALGEBRAIC MANIPULATION

Once you have written down an algebraic expression, like $1{,}000a + 100b + 10c + d$, what are you going to do with it? As it stands it has not given us any new information. Something always has to be done to the expression: either we must combine it with other expressions, as we combined a number with 11 in multiplication a few paragraphs back, or else we have to change it in some other way. Whatever we do, it will come under the formal and rather dignified name of "algebraic manipulation."

Undoubtedly you have encountered this sort of thing in school, either as algebra or as arithmetic. Perhaps part of it was only suggested by examples, and never clearly stated, but it was there in some way. But some of the possible kinds of manipulation may not be fresh in your mind, and perhaps also you may be tempted to try some incorrect kinds of manipulation. Certain rearrangements of the numbers and letters look plausible, but actually would lead to wrong answers. So to refresh your memory, let us list the legitimate ways of changing algebraic expressions:

Grouping in parentheses or brackets

We did this in the "rule for eleven" just above, because we had $(a + b) \times 1{,}000$ and similar expressions. Take the case where a happens to be 2 and b happens to be 3; then $a + b$ is 5. In this case the expression $(a + b) \times 1{,}000$ is 5,000. This is a natural and convenient way to proceed.

We have to be a little careful, though. In complicated expressions there is some danger of making an error unless we either remember certain rules, or understand the basic ideas very thoroughly.

There is really only one truly basic idea in the use of parentheses, or brackets. They tell us to think of everything inside as one number. This is in line with what the written symbols naturally suggest. Suppose we had $2 \times (5 + 1)$, for instance. We wish to think of $5 + 1$ as a single idea, so it is held together by parentheses. Then we replace $(5 + 1)$ by the corresponding single number, 6, and we have

$$\begin{aligned} 2 \times (5 + 1) &= 2 \times 6 \\ &= 12 \end{aligned}$$

Suppose we had a subtraction inside brackets or parentheses, like $2 \times (5 - 1)$. We use the same principle, and we have

$$\begin{aligned} 2 \times (5 - 1) &= 2 \times 4 \\ &= 8 \end{aligned}$$

We use the word "parenthesis" for the curved symbol and the word "bracket" for the square one. A common situation is to have parentheses within brackets, like this:

$$2 \times [(5 + 1) - (3 - 2)]$$

What shall we do here? Well, the principle is that we wish

to think of everything inside an enclosure as one number. The result is that we must start with the innermost enclosure. That is, we cannot do anything immediately with $[(5 + 1) - (3 - 2)]$, because we do not know immediately what the figures inside of the square brackets amount to. The place to start is with something that we do know immediately. This is $5 + 1$, which we can replace with 6. Likewise we know that $3 - 2$ is 1. So we can certainly write

$$2 \times [(5 + 1) - (3 - 2)] = 2 \times [6 - 1]$$

Then we are almost finished. Because $6 - 1$ is 5 and so

$$\begin{aligned} 2 \times [(5 + 1) - (3 - 2)] &= 2 \times [6 - 1] \\ &= 2 \times 5 \\ &= 10 \end{aligned}$$

This leads us to the rule: *Start with the innermost and work out.*

With letters instead of numbers it is a little different, simply because it is obviously impossible to get rid of parentheses by actual calculation. For instance, $(a + b) \times 1{,}000$ cannot be simplified by actually performing the addition, because we do not wish to give any particular values to a and b. In this case, we would probably leave it in the form it has.

Frequently, though, we find it convenient to get rid of the parentheses by another method. This is what is called "removing parentheses," and in this example it would go like this:

$$2 \times (5 + 1) = 2 \times 5 + 2 \times 1$$

We have omitted the parentheses, and to compensate for the omission we have applied the idea of multiplying everything inside the parentheses separately by 2. Notice that this gives the correct result:

$$\begin{aligned} 2 \times (5 + 1) &= 2 \times 5 + 2 \times 1 \\ &= 10 + 2 \\ &= 12 \end{aligned}$$

The result, 12, is the same as what we found before from 2 times 6.

In terms of letters we might have something like this:

$$a(x + y + z) = ax + ay + az$$

Here we have multiplied by the number a. On the left side of the equal sign we multiplied the sum of $x + y + z$ by a. On the right we multiplied each of the three by a separately. If $a = 3$, and $x = 5$, $y = 2$, and $z = 4$ (we took these out of the air, just as an illustration), we would have:

$$3(5 + 2 + 4) = 3 \times 5 + 3 \times 2 + 3 \times 4$$

which is $\quad 3 \times 11 = 15 + 6 + 12$

or $\quad 33 = 33$

So the method, applying the multiplier to each term separately, works out correctly here, as it must always do.

When only addition is involved we simply remove the parentheses:

$$2 + (5 + 1) = 2 + 5 + 1$$

which is, $\quad 2 + 6 = 7 + 1$

or $\quad = 8$

A subtraction *inside* of the enclosure causes no trouble. Again we merely omit the parentheses:

$$2 + (5 - 1) = 2 + 4$$

that is, $\quad 2 + 4 = 6$

But a minus sign *outside* the enclosure, in front of the whole enclosure, tells us to subtract everything inside the parentheses as if it were a single number, and this causes a little trouble. When we remove the parentheses, we must reverse all the signs inside the enclosure. Every plus sign becomes minus and every minus becomes plus. Like this:

$$8 - (5 - 1 + 3 - 2) = 8 - 5 + 1 - 3 + 2$$

On the left we have two minuses, -1 and -2, and a plus, $+3$. The 5 is also understood to have a plus sign. *Whenever a number is written without any sign before it, neither plus nor minus, it is understood that the sign is plus.*

Notice that this works out because on the left side of the equation:

$$8 - (5 - 1 + 3 - 2) \text{ becomes } 8 - (4 + 1) = 8 - 5 = 3$$

And on the right:

$$8 - 5 + 1 - 3 + 2 \text{ becomes } 3 + 1 - 3 + 2 = 3$$

The left side of the equation equals 3, and the right side also equals 3, so the equation is true.

The same kind of situation in terms of letters would be like this:

$$a - (m - n + s - t) = a - m + n - s + t$$

Notice that we can do the same thing in reverse if we wish. Instead of removing parentheses we can put them in where none were before, and sometimes it is to our advantage to do so. Whether we remove parentheses or put new ones in depends on the specific situation. We know that:

$$2(a + b + c) = 2a + 2b + 2c$$

The two expressions, one on the left and the other on the

right of the equal sign, are equal to each other, and we can replace either one by the other in a problem or calculation. So if we happened to notice, in the course of working a problem, that we had the expression $2a + 2b + 2c$, we would have the right to replace it by $2(a + b + c)$ if we wished. We can think of this as "extracting" the 2. This is very often useful.

Frequently we encounter two expressions in parentheses together, in the form of

ADDITION: $(a + d) + (c - d)$
OR SUBTRACTION: $(a + b) - (c - d)$
OR MULTIPLICATION: $(a + b)(c - d)$

The last one means exactly the same as writing $(a + b) \times (c - d)$, but it is customary to omit the times sign when we have letters instead of numbers. In all these cases the basic idea is to take two steps:

1. *Remove one pair of parentheses—either one—leaving the other unchanged, that is, we open up only one of the two enclosures; then*
2. *Remove the second pair of parentheses.*

For instance, in a simple addition we would have:

ADDITION:

$$(a + b) + (c - d) = a + b + (c - d)$$
$$= a + b + c - d$$

still thinking of $(c - d)$ as one quantity!

SUBTRACTION:

$$(a + b) - (c - d) = a + b - (c - d)$$
$$= a + b - c + d$$

d changes its sign

MULTIPLICATION:

$$(a + b) \times (c - d) = (a + b)(c - d)$$
$$= a(c - d) + b(c - d)$$
$$= ac - ad + bc - bd$$

still thinking of $(c - d)$ as one number

Equations

In order to manipulate equations, we make use of one basic principle under several different forms. It is essentially this: whatever expression is written on the left of the equal sign is one way of writing a certain quantity, and whatever is written on the right is a different way of writing the *same* quantity. For instance:

$$a + 2b - 1 = 15$$

means that $a + 2b - 1$ is one way of writing the quantity 15. Anything that we do to $a + 2b - 1$, such as doubling it, or adding 1 to it, must also be done to the 15 on the right of the equal sign, in order to maintain the equality:

	$a + 2b - 1 = 15$
DOUBLE:	$2(a + 2b - 1) = 30$
ADD 1:	$a + 2b - 1 + 1 = 16$
SQUARE:	$(a + 2b - 1)(a + 2b - 1) = 15 \times 15$

In a few words: *Anything that we do to the left-hand side of any equation must also be done to its right-hand side.* But always remember this: everything to the left of the equal sign must be treated as a single quantity, and so must the expression on the right. That is, it must be treated as if it were in parentheses, as we did by actually writing the parentheses in the doubled and the squared equations above.

The ideas of the preceding sections also enable us to do a little further manipulating, such as:

THE DOUBLED EQUATION:	$2a + 4b - 2 = 30$
THE "ADD 1" EQUATION:	$a + 2b = 16$
THE SQUARED EQUATION:	$(a + 2b - 1)^2 = 225$

The expression in the last equation, with the figure two

written above the line, is something that we had before in the chapter on squares and square roots. The ^{2} is read as the word "squared," and it means that the expression to which it is attached is multiplied by itself. The expression 7^2 means 49, because it means 7 multiplied by itself. We write ^{2} because we have *two* sevens multiplied together.

The basic principle of doing the same thing to both sides of the equation appears in several forms, according to what the "same thing" happens to be. Two forms are particularly useful:

1. Adding the same number to both sides of the equation. This includes subtracting the same number from both sides, because subtracting is the same as adding a negative number. For instance:

$$x - 1 = 5$$

Add 1 to both sides, and we have

$$x - 1 + 1 = 5 + 1$$

that is,

$$x = 6$$

This is often referred to as "transposing a number to the other side." We have in effect transposed, or moved, the 1 of $x - 1$ to the right-hand side, and added it to the 5. On the left it was minus 1, and on the right it is plus 1. This always happens. You may think of it as a rule, in fact it is often stated as a rule: *When you transpose, change the sign.*

Once you understand what happened in the example, when $x - 1$ became $x - 1 + 1$ which is $x + 0$, you will have the rule in your possession without the effort of memorizing it. For what happened was, we added to the left side what we needed to wipe out the minus 1—we added 1—cancelling the minus 1 on the left side. To preserve equality we must also add 1 to the right-hand side, and we did so. Any term, in any equation, can be transposed to the other side of the equation in the same way.

2. Multiplying or dividing both sides of the equation by the same number. For instance:

$$3 + 4 = 7$$

Multiply by 5: $5(3 + 4) = 35$

that is, either: $5 \times 7 = 35$

or: $15 + 20 = 35$

In algebraic usage, when letters are used instead of numbers, we have the same process, as in this example:

$$x^2 + x + \tfrac{3}{4} = \tfrac{7}{4}$$

Multiplied by 4: $4(x^2 + x + \tfrac{3}{4}) = 4 \times \tfrac{7}{4} = 7$

that is: $4x^2 + 4x + 3 = 7$

This has simplified the form of the equation by eliminating the fractions. The new equation is equivalent to the one with fractions, but it will be easier to deal with in later manipulations.

Example 1: We had a puzzle at the beginning of this chapter. Let us solve it directly, by using algebra. It consisted of three numbers—the lengths of three boards, but we need only think of them as numbers—and these numbers were described as having the following properties:

(1) The first number is known to be 3.
(2) The second one is equal to the first plus one-fourth of the third,
(3) The third is equal to the first two added together.

Let us call the first of these numbers x, the second one y, and the third one z. The three statements listed just above can be written in this way:

$$(1)\ x = 3$$
$$(2)\ y = x + \tfrac{1}{4}z$$
$$(3)\ z = x + y$$

Use equation (1) $x = 3$, to get rid of x, wherever we encounter it in the other two equations:

$$(2)\ y = 3 + \tfrac{1}{4}z$$
$$(3)\ z = 3 + y$$

Now this new form of (2) makes it possible for us to eliminate y from (3). We replace y in (3):

$$(3)\ z = 3 + (3 + \tfrac{1}{4}z)$$
$$= 6 + \tfrac{1}{4}z$$

This contains fractions, and fractions are less convenient to work with than whole numbers, so we get rid of the fractions. Multiply by 4:

$$4z = 4 \times 6 + 4 \times \tfrac{1}{4}z$$
$$4z = 24 + z$$

Subtract z from both sides (or "transpose" the z):

$$4z - z = 24 + z - z$$
$$3z = 24$$

Divide both sides by 3, and we have part of the answer:

$$z = 8$$

What is y? We can find it now by using equation (3):

$$z = 3 + y$$

becomes: $8 = 3 + y$

Subtract 3 from both sides:

$$5 = y$$

that is: $$y = 5$$

What is x? We already know that x is 3, because in this case it happened to be given to us. In other examples, we could find out what x was by using the values of z and y which we have just found. The answer is:

$$x = 3$$
$$y = 5$$
$$\underline{z = 8}$$

Total length of board, 16.

Example 2: This is adapted from an ancient Persian book on mathematics:

A lady of the court was wearing a pearl necklace one night. In an amorous struggle, the necklace was broken, and one-third of the pearls fell on the floor. A fourth of the pearls remained on the couch, and there were twenty pearls remaining on the string. How many pearls were on the string originally?

We let x represent the number of pearls on the string before it was broken. Total number of pearls:

BEFORE THE ACCIDENT $= x$ *all on the string*

AFTER THE ACCIDENT $= \frac{1}{3}x + \frac{1}{4}x + 20$ *a third on the floor,*
a fourth on the couch,
20 remained

But these must be equal, before and after the accident, because all the pearls are accounted for. We write the equality:

$$x = \tfrac{1}{3}x + \tfrac{1}{4}x + 20$$

Eliminate fractions by multiplying through by 12:

$$12x = 12 \times \tfrac{1}{3}x + 12 \times \tfrac{1}{4}x + 12 \times 20$$

that is: $$12x = 4x + 3x + 240 = 7x + 240$$

Subtract $7x$ from both sides of the equation:

$$5x = 240$$

that is: $$x = 48$$

The necklace was one of 48 pearls. You can easily verify that a necklace of 48 pearls fits the description given in the problem.

THE TRACHTENBERG SYSTEM IN ALGEBRA

The Rule for Six

We shall use these methods of manipulating equations to show that the rule for six, and other parts of the system, actually do give the correct answer. This shows at the same time *how* the rule arrives at the correct answer, which is interesting because it gives us insight into the situation.

The rule for six tells us to "add half the neighbor, plus five if odd," meaning if the number itself is odd, not if the neighbor is odd. This has the effect of multiplying by six. To arrive at the rule we write six in a very special way:

$$6 = 5 + 1$$
$$6 = \tfrac{1}{2} \times 10 + 1$$

This is the way we write 6. How shall we write the number that is to be multiplied by 6? We can call this number N if we wish, and we choose to write it in this way (like 4028; we do not wish to restrict the first digit to being 4, so we simply call it a, and similarly with the other figures, b, c, d):

$$N = a\ b\ c\ d$$
$$N = a \times 1{,}000 + b \times 100 + c \times 10 + d$$
$$= 1{,}000a + 100b + 10c + d$$

We are showing N as being a four-figure number, but this is only for the sake of definiteness. A five-figure number would begin with 10,000 times a, and so on.

We wish to multiply this number N by 6. To avoid any possible confusion, let us indicate multiplication by writing a dot between the numbers multiplied, like $5 \cdot 7$ for 5 times 7. Multiply the long number by 6:

$$6 \cdot N = (\tfrac{1}{2} \cdot 10 + 1) \cdot N \qquad \textit{because 6 equals } \tfrac{1}{2} \cdot 10 + 1$$
$$= \tfrac{1}{2} \cdot 10 \cdot N + N \qquad \textit{by removing parentheses}$$

Replace the four-digit number N by its full form:

$$6 \cdot N = \tfrac{1}{2} \cdot 10 \cdot (a \cdot 1{,}000 + b \cdot 100 + c \cdot 10 + d)$$
$$+ 1 \cdot (a \cdot 1{,}000 + b \cdot 100 + c \cdot 10 + d)$$

Remove the parentheses in this equation, both pairs:

$$6 \cdot N = \tfrac{1}{2} \cdot 10 \cdot a \cdot 1{,}000 + \tfrac{1}{2} \cdot 10 \cdot b \cdot 100 + \tfrac{1}{2} \cdot 10 \cdot c \cdot 10$$
$$+ \tfrac{1}{2} \cdot 10 \cdot d + a \cdot 1{,}000 + b \cdot 100 + c \cdot 10 + d$$

In the first term after the equal sign, we can multiply the 10 and the 1,000 together to get 10,000, so that:

$$\tfrac{1}{2} \cdot 10 \cdot a \cdot 1{,}000 \text{ becomes } \tfrac{1}{2} \cdot a \cdot 10{,}000$$

We can do similar multiplications in other terms. The result is this:

$$6 \cdot N = \tfrac{1}{2} \cdot a \cdot 10{,}000 + \tfrac{1}{2} \cdot b \cdot 1{,}000 + \tfrac{1}{2} \cdot c \cdot 100 + \tfrac{1}{2} \cdot d \cdot 10$$
$$+ a \cdot 1{,}000 + b \cdot 100 + c \cdot 10 + d$$

Now we come to the crucial point. We rearrange these terms. The two terms of the form "something times 1,000" are put together. Those times 100 are put together, and so on:

$$\begin{aligned} 6 \cdot N = {} & \tfrac{1}{2} \cdot a \cdot 10{,}000 \\ & + \tfrac{1}{2} \cdot b \cdot 1{,}000 + a \cdot 1{,}000 \\ & + \tfrac{1}{2} \cdot c \cdot 100 + b \cdot 100 \\ & + \tfrac{1}{2} \cdot d \cdot 10 + c \cdot 10 \\ & + d \end{aligned}$$

Now we can insert parentheses, as we did a few pages back. The second line just above shows that two terms are added together, and each of these terms is something times a thousand. We can "factor out" the 1,000, and what remains goes into parentheses:

$$\tfrac{1}{2} \cdot b \cdot 1{,}000 + a \cdot 1{,}000 = (\tfrac{1}{2} \cdot b + a) \cdot 1{,}000$$

The other lines behave similarly:

$$\begin{aligned} 6 \cdot N = {} & \tfrac{1}{2} \cdot a \cdot 10{,}000 \\ & + (a + \tfrac{1}{2} \cdot b) \cdot 1{,}000 \\ & + (b + \tfrac{1}{2} \cdot c) \cdot 100 \\ & + (c + \tfrac{1}{2} \cdot d) \cdot 10 \\ & + (d + \tfrac{1}{2} \cdot 0) \cdot 1 \qquad \textit{because anything times zero equals zero} \end{aligned}$$

The pattern here is obvious. We have added the $\frac{1}{2} \cdot 0$ term for the sake of completing the pattern; it is permissible to add zero to anything if we wish, since adding zero does not increase or decrease the number.

The pattern is still not quite complete. We can complete it by adding a zero term at the beginning, on the first line. We write $\frac{1}{2} \cdot a \cdot 10{,}000$ in the form $0 + \frac{1}{2} \cdot a \cdot 10{,}000$. Then:

$$\begin{aligned} 6 \cdot N = {} & (0 + \tfrac{1}{2} \cdot a) \cdot 10{,}000 \\ & + (a + \tfrac{1}{2} \cdot b) \cdot 1{,}000 \\ & + (b + \tfrac{1}{2} \cdot c) \cdot 100 \\ & + (c + \tfrac{1}{2} \cdot d) \cdot 10 \\ & + (d + \tfrac{1}{2} \cdot 0) \cdot 1 \end{aligned}$$

This demonstrates the "rule for six." The original number N had the form of a four-digit number a,bcd when written in the ordinary way, so that a is the figure in the thousands position and so on. Under the a, in the thousands position of the answer, we have $(a + \frac{1}{2} \cdot b) \cdot 1{,}000$, that is, the given figure in that position plus half of its right-hand neighbor $(\frac{1}{2} \cdot b)$. Similarly in all other positions, we "add half the neighbor."

This is the whole story if the given number N contains only even digits, like 2 and 6. What if there is a 3 or a 7? The rule says then "add half the neighbor plus 5 if the figure is odd." In that case one of the expressions $\frac{1}{2} \cdot a$, $\frac{1}{2} \cdot b$, etc., would be fractional, because it would be half of 7 or something similar.

We take care of the possibility in this way. We assume that one of the figures, say b for instance, is odd, and we replace it by the expression $2n + 1$. Any *odd* whole number can be written in this form. For instance, 7 is $2 \cdot 3 + 1$, and 9 is $2 \cdot 4 + 1$. The number n here is what we have called "the smaller half" of an odd figure. Replace b by $2n + 1$ in the equations given above, and you will find that the same argument holds up to the end. We have:

$$\begin{aligned} 6 \cdot N = {} & (0 + \tfrac{1}{2} \cdot a) \cdot 10{,}000 \\ & + (a + n + \tfrac{1}{2}) \cdot 1{,}000 \\ & + (2n + 1 + \tfrac{1}{2} \cdot c) \cdot 100 \\ & + etc. \end{aligned}$$

$$\begin{aligned} = {} & (0 + \tfrac{1}{2}a) \cdot 10{,}000 + (a + n) \cdot 1{,}000 + 500 \\ & + (2n + 1 + \tfrac{1}{2} \cdot c) \cdot 100 \\ & + etc. \end{aligned}$$

$$= (0 + \tfrac{1}{2}a) \cdot 10{,}000 + (a + n) \cdot 1{,}000$$
$$+ (b + \tfrac{1}{2} \cdot c + 5) \cdot 100 \quad \text{(replacing } 2n + 1 \text{ by } b\text{)}$$
$$+ \textit{etc.}$$

This tells us to use the "smaller half" and to add the 5, as the rule has already told us to do. The rule is verified.

This "rule for six" argument was shown at full length for the sake of clarity. We do not need to repeat this procedure —the method of arriving at the Trachtenberg rules is similar for the other rules as far as the techniques are concerned. It will be sufficient to present a few of the other parts of the method in a more abbreviated form, if this rule for six has been understood.

NON-TABLE MULTIPLICATION IN GENERAL

The other "rules" are derived by arguments which run along the same general lines as the preceding section. This is true in a general sense only, because the rules for eight and nine are somewhat different. To summarize:

1. The "rule for seven" is very similar to the rule for six, requiring only the doubling of the number. In consequence, the derivation of the rule for seven is the same as that for six in the preceding section, except that we use $7 \cdot N$ instead of $6 \cdot N$, and we replace the 7 by $\frac{1}{2} \cdot 10 + 2$, just as we replaced the 6 by $\frac{1}{2} \cdot 10 + 1$.

2. The "rule for five" is again along the same lines, in fact exactly the same except for one thing. Instead of having $6 \cdot N = (\frac{1}{2} \cdot 10 + 1) \cdot N$ we now have the simpler form, $5 \cdot N = \frac{1}{2} \cdot 10 \cdot N$.

Both the rule for seven and that for five can be easily worked out by following along the argument of the preceding section and making the appropriate changes, if you wish to do so for your own edification or amusement.

3. The rules for nine and eight both require a different approach, or rather a different "trick" at one point, as we shall now see.

The rule for nine

Multiplying by 9 without using multiplication tables, we follow this procedure (as you perhaps remember):

(1) Subtract the right-hand figure from 10.

(2) Subtract each other figure from 9 and add the neighbor.

(3) When you come to the final figure, the one at the left-hand end of the answer, we use the left-hand figure of the given number minus 1.

Throughout all this it is understood that we "carry" a 1 if necessary, as usual (nothing higher than 1 will turn up).

Let us see where this rule comes from. We can write 9 in the form $10 - 1$ if we wish. We do wish, because this will lead us to the rule. For any number a, it is also true that $9a$, or 9 times a, is equal to $10a - a$.

The number that we wish to multiply by 9 may be called N if we wish, and may be expanded into a fuller form as we did before:

$$\begin{aligned} 9 \cdot N &= 9 \cdot (a \cdot 1{,}000 + b \cdot 100 + c \cdot 10 + d) \\ &= 9 \cdot a \cdot 1{,}000 + 9 \cdot b \cdot 100 + 9 \cdot c \cdot 10 + 9 \cdot d \end{aligned}$$

Now use the fact that 9 is 10 minus 1, and $9a = 10a - a$, etc:

$$\begin{aligned} 9 \cdot N &= 10 \cdot a \cdot 1{,}000 - a \cdot 1{,}000 + 10 \cdot b \cdot 100 - b \cdot 100 \\ &\quad + 10 \cdot c \cdot 10 - c \cdot 10 + 10 \cdot d \cdot 1 - d \cdot 1 \\ &= a \cdot 10{,}000 - a \cdot 1{,}000 + b \cdot 1{,}000 - b \cdot 100 + c \cdot 100 \\ &\quad - c \cdot 10 + d \cdot 10 - d \end{aligned}$$

This is similar to what we did in the preceding section, dealing with the rule for six.

Here is the point at which we need a new device, to handle the rule for nine. We add and subtract the same number, which is legitimate because it does not change the quantities. For instance, if we start with 25 and then we both add and subtract 2, we have $25 + 2 - 2$, which is still 25. Adding and subtracting the same number means that we are adding zero, which does not change the size of a number: 25 plus zero is still 25. So we are justified in writing 25 as $25 + 2 - 2$, or as $25 + 7 - 7$, or anything similar, if we wish to do so.

Does this seem to be pointless? Such an impression would be false. It can be helpful whenever we have several terms added together. In such cases it is possible to group the *subtracted* term, like -2, with one group of the other terms, and then group the *added* term, like the $+2$, with a different set of terms. This grouping will sometimes, when we are lucky, result in simplifying both groups of terms.

In this example, the rule for nine, we add and subtract 9,000, also 900, also 90, and 9, like this:

$$\begin{aligned} 9 \cdot N = a \cdot 10{,}000 &- 9{,}000 + 9{,}000 - a \cdot 1{,}000 + b \cdot 1{,}000 \\ &- 900 + 900 - b \cdot 100 + c \cdot 100 - 90 + 90 \\ &- c \cdot 10 + d \cdot 10 - 9 + 9 - d \end{aligned}$$

Group together the "thousands" terms, the "hundreds" terms, and so on, and this becomes:

$$\begin{aligned} 9 \cdot N = a \cdot (10{,}000) &+ (9 - a + b) \cdot 1{,}000 + (9 - b + c) \cdot 100 \\ &+ (9 - c + d) \cdot 10 + (9 - d) \cdot 1 \\ &- (9{,}000 + 900 + 90 + 9) \end{aligned}$$

The numbers in the parenthesis on the last line amount to 9,999, which we write as $10{,}000 - 1$. Then:

$$\begin{aligned}9 \cdot N &= a \cdot 10{,}000 + (9 - a + b) \cdot 1{,}000 + (9 - b + c) \cdot 100 \\ &\quad + (9 - c + d) \cdot 10 + (9 - d) \cdot 1 - 10{,}000 + 1 \\ &= (a - 1) \cdot 10{,}000 + (9 - a + b) \cdot 1{,}000 \\ &\quad + (9 - b + c) \cdot 100 + (9 - c + d) \cdot 10 + (10 - d) \cdot 1\end{aligned}$$

This is precisely the "rule for nine," expressed in symbols.

The Rule for Eight

Write 8 as 10 − 2, which is similar to writing 9 as 10 − 1. Then follow along the same lines as in the preceding section, except for a slight change when we come to the business of adding and subtracting the same numbers. Before, we added and subtracted 9,000, 900, 90, and 9. Now, in the rule for eight, we add and subtract double these figures; that is, 18,000, 1,800, 180, and 18. The result is that we have to double what we get by subtracting from 9 (or from 10, at the first step), and the left-hand figure is 2 less than the left-hand figure of the given number, not 1 less. This is precisely the rule for eight.

SQUARING NUMBERS

In a previous chapter we had special methods for finding the square of a number, that is, for multiplying a number by itself. We began with two particularly interesting kinds of numbers:

1. Two-digit numbers, of which the second digit is 5, like 35. To multiply 35 by 35, we multiply 3 by the next larger digit, 4, and we have 12. Write 25 after this 12, and you have the answer, 1,225.

In algebraic symbols, such numbers have the form $a \cdot 10 + 5$.

The desired result, the square, is $(a \cdot 10 + 5)^2 = (a \cdot 10 + 5)(a \cdot 10 + 5)$.

Expand the parentheses:

$$\begin{aligned}(a \cdot 10 + 5) \cdot (a \cdot 10 + 5) &= a \cdot 10 \cdot (a \cdot 10 + 5) + 5 \cdot (a \cdot 10 + 5)\\ &= a \cdot a \cdot 100 + a \cdot 50 + a \cdot 50 + 25\\ &= a \cdot a \cdot 100 + a \cdot 100 + 25\end{aligned}$$

Now we group the first two terms together and we can take out the factor a by inserting parentheses, as we did earlier in this chapter. The expression now becomes:

$$\begin{aligned}(a \cdot 10 + 5)^2 &= a \cdot (a \cdot 100 + 100) + 25\\ &= a \cdot (a + 1) \cdot 100 + 25\end{aligned}$$

This is the rule, in symbols. For $a(a + 1)$ is the original tens-digit times a digit larger by 1, and the factor 100 tells us that the result cannot overlap the 25 (because multiplying a number by 100 has the effect of placing two zeroes after it).

2. If the first digit of a two-digit number is 5, as is 56, we square the 5 to get 25, and we add to this the units-digit (the 6, in the case of 56). The result is the first two digits of the answer: in the case of 56, we have the partial answer $56^2 = 31??$ To fill in the last two digits we simply square the units-digit of the given number; in the case of 56 we square the 6 and we have 36. This 36 is the rest of the answer, and the whole number is 3,136.

Such a number has the form $(5 \cdot 10 + a)^2$, which is $(5 \cdot 10 + a) \cdot (5 \cdot 10 + a)$. Expand the parentheses as we did above:

$$\begin{aligned}(5 \cdot 10 + a)^2 &= 5 \cdot 10 \cdot 5 \cdot 10 + 5 \cdot 10 \cdot a + 5 \cdot 10 \cdot a + a \cdot a\\ &= 25 \cdot 100 + 100 \cdot a + a^2\\ &= (25 + a) \cdot 100 + a^2\end{aligned}$$

This is equivalent, in symbols, to the procedure described in the first paragraph of this section, where we squared 56.

3. To square any two-digit number at all, say for instance 73, we find the:

(1) units-digit of the answer by squaring the given units,
(2) tens-digit of the answer by doubling the cross-product of the given number (for 73, we double 7 times 3 and we have 42), and
(3) hundreds- and thousands-digits of the answer by squaring the tens-digit of the given number.

As with 73:

$$\frac{\quad 7\ 3\ ^{2}}{5\ 3\ {}^{4}2\ 9}$$

The general form of such a number is $a \cdot 10 + b$. Square it:

$$\begin{aligned}(a \cdot 10 + b)^2 &= (a \cdot 10 + b) \cdot (a \cdot 10 + b) \\ &= a \cdot 10 \cdot (a \cdot 10 + b) + b \cdot (a \cdot 10 + b) \\ &= a \cdot 10 \cdot a \cdot 10 + a \cdot 10 \cdot b + b \cdot a \cdot 10 + b^2 \\ &= a^2 \cdot 100 + 2a \cdot b \cdot 10 + b^2\end{aligned}$$

This corresponds to the procedure stated just above. The product $a \cdot b$ is the cross-product, tens-digit times units-digit, and the equation shows that we must double it.

MULTIPLICATION BY THE UNITS-AND-TENS METHOD

First let us consider the multiplication of a three-digit number by a one-digit number, like 617 times 3:

$$\frac{0\ 6\ 1\ 7}{1\ 8\ 5\ 1} \times 3$$

This is done, as you will recall, by moving a pattern UT across the 617 from right to left:

```
      U T
0 6 1 7   ×   3
-------
      1
```

1 is the "units" of 21, i.e. 7 times 3

and then

```
    U T
0 6 1 7   ×   3
-------
    5
```

5 equal units of 1 times 3 plus tens of 7 times 3

And so on.

Let us look at the case of any three-digit number, in algebraic terms. Such a number has the appearance, in ordinary writing, of *a b c,* and is written in full as:

$$a \cdot 100 + b \cdot 10 + c \cdot 1$$

Let us call our multiplier *n*. It is a single figure. Then:

$$(a \cdot 100 + b \cdot 10 + c \cdot 1) \cdot n$$

is our multiplication in algebraic form. Expand the bracket:

$$(a \cdot 100 + b \cdot 10 + c \cdot 1) \cdot n$$
$$= n \cdot a \cdot 100 + n \cdot b \cdot 10 + n \cdot c \cdot 1$$

Each of the pairs $n \cdot a$, $n \cdot b$, and $n \cdot c$ is the product of two single figures. The result is in general a two-figure number; for instance, 7 times 3 is 21, a two-figure number. To keep matters straight we must write them in the form of two-digit numbers. We can do so by introducing new symbols with subscripts. One of them is U_a, which means a single figure, namely the units-digit of what we get when we multiply a by the multiplier n. The multiplier n is the given multiplier for the problem, and so we need not mention it in the symbol. The subscript a of U_a tells us to multiply our n by a, and the U of U_a tells us to take the units of the result. Similarly for the others:

$$n \cdot a = T_a \cdot 10 + U_a \qquad \textit{tens and units of } a \textit{ times } n$$
$$n \cdot b = T_b \cdot 10 + U_b$$
$$n \cdot c = T_c \cdot 10 + U_c$$

Then our desired multiplication now looks like this:

$$(a \cdot 100 + b \cdot 10 + c \cdot 1) \quad \times \quad n$$
$$= (T_a \cdot 10 + U_a) \cdot 100 + (T_b \cdot 10 + U_b) \cdot 10 + (T_c \cdot 10 + U_c) \cdot 1$$

Expand the parentheses:

$$(a \cdot 100 + b \cdot 10 + c \cdot 1) \quad \times \quad n$$
$$= T_a \cdot 1{,}000 + U_a \cdot 100 + T_b \cdot 100 + U_b \cdot 10 + T_c \cdot 10 + U_c$$
$$= T_a \cdot 1{,}000 + (U_a + T_b) \cdot 100 + (U_b + T_c) \cdot 10 + U_c$$

Remembering the facts that:

(a) the letters T and U mean the units- and tens-digits of what we get by multiplying by the given multiplier n, and that

(b) the subscript tells us which digit of the long number is being multiplied by n,

we see that this last equation describes the method of units-and-tens multiplication. Take for instance the term $(U_b + T_c) \cdot 10$:

$$U_b = \text{units of } b \text{ times the multiplier, } n$$
$$T_c = \text{tens of } c \text{ times } n$$

Then apply this to the multiplication written in our usual way:

$$\underline{a \quad b \quad c} \quad \times \quad n$$

Place the U_b and the T_c above the digits that they refer to:

$$\begin{array}{ccccc} & U_b & T_c & & \\ \hline a & b & c & \times & n \end{array}$$

Once they are correctly placed, we do not need the subscripts. We can write it simply as:

$$\begin{array}{ccccc} & U & T & & \\ a & b & c & \times & n \\ \hline & * & & & \end{array}$$

this gives a figure of the answer where the asterisk is

The other figures of the answer come out of the other terms of the equation, in exactly the same way. Thus we have arrived at the method of units-and-tens multiplication, for the case of multiplying by a single digit.

Multiplication by longer multipliers

Suppose we wish to multiply 617 by 23. We use two UT patterns to find each figure of the answer. We obtain the figure of the answer by adding together the two numbers that result from the two UT patterns, like this:

$$\begin{array}{cccccccc} & & U & T & & & & \\ & & & U & T & & & \\ 0 & 0 & 6 & 1 & 7 & \times & 2 & 3 \\ \hline & & & \dot{}1 & & & & \end{array}$$

11; 61 with 3 gives $1\underline{8} + \underline{0}3 = 8$;
17 with 2 gives $0\underline{2} + \underline{1}4 = 3$ and $8 + 3 = 11$

This example worked out in full is

$$\begin{array}{l} \underline{0\ 0\ 6\ 1\ 7\ } \times\ 2\ 3 \\ 1\ 4\dot{}1\ 9\ 1 \end{array}$$

Suppose we have any three-digit number, say *abc,* to be multiplied by any two-digit multiplier, *mn.* When written out in full the multiplication is:

$$(a\cdot 100 + b\cdot 10 + c\cdot 1) \quad \times \quad (m\cdot 10 + n)$$

Expand the parentheses: the desired answer is equal to

$$\begin{aligned} & a\cdot 100\cdot m\cdot 10 + a\cdot 100\cdot n + b\cdot 10\cdot m\cdot 10 + b\cdot 10\cdot n \\ & \qquad + c\cdot m\cdot 10 + c\cdot n \\ & = a\cdot m\cdot 1{,}000 + a\cdot n\cdot 100 + b\cdot m\cdot 100 + b\cdot n\cdot 10 \\ & \qquad + c\cdot m\cdot 10 + c\cdot n\cdot 1 \end{aligned}$$

In the first term we see $a\cdot m$, which is two single digits multiplied together. This is in general a two-digit number. We must write out all these two-digit numbers, $a\cdot m$, $a\cdot n$, $b\cdot m$, and so on, in the form of two-digit numbers. This was done just above, when we had a single-digit multiplier. We did it by introducing the symbols T_a, U_a, and so on.

But now we need another subscript, because the multiplier involved is not always the same digit. Half the time it is *m* and half the time it is *n.* In the example above, where we multiplied by 23, half the time we acted upon a pair with the 2 and half the time with the 3, of the 23. To remind ourselves which one is being used at the moment we must write another subscript, and we have such symbols as U_{am}. This is still only a single digit, even though three letters are used in writing it; we write the two subscripts only for our own convenience, to remind ourselves what numbers are being multiplied.

These are the two-digit expressions that we need:

$$\begin{aligned}
a \cdot m &= T_{am} \cdot 10 + U_{am} \\
a \cdot n &= T_{an} \cdot 10 + U_{an} \\
b \cdot m &= T_{bm} \cdot 10 + U_{bm} \\
b \cdot n &= T_{bn} \cdot 10 + U_{bn} \\
c \cdot m &= T_{cm} \cdot 10 + U_{cm} \\
c \cdot n &= T_{cn} \cdot 10 + U_{cn}
\end{aligned}$$

The desired answer of our multiplication now takes the form:

$$\begin{aligned}
&(T_{am} \cdot 10 + U_{am}) \cdot 1{,}000 + (T_{an} \cdot 10 + U_{an}) \cdot 100 \\
&(T_{bm} \cdot 10 + U_{bm}) \cdot 100 + (T_{bn} \cdot 10 + U_{bn}) \cdot 10 \\
&(T_{cm} \cdot 10 + U_{cm}) \cdot 10 + (T_{cn} \cdot 10 + U_{cn})
\end{aligned}$$

Now expand these parentheses and rearrange the terms, and we have the final result:

$$\begin{aligned}
\text{ANSWER} = T_{am} \cdot 10{,}000 &+ (T_{an} + U_{am} + T_{bm}) \cdot 1{,}000 \\
&+ (U_{an} + T_{bn} + U_{bm} + T_{cm}) \cdot 100 \\
&+ (U_{bn} + T_{cn} + U_{cm}) \cdot 10 + U_{cn}
\end{aligned}$$

This is the statement, in terms of algebraic symbols, of the method of adding together the results of our two UT pairs at each step of the multiplication. The same method of proof will apply to the case of longer numbers times longer multipliers than the ones shown.

NUMBERS OF ANY LENGTH

In the last few sections we have been handling numbers of a general form, such as the number $a \cdot 1{,}000 + b \cdot 100 + c \cdot 10 + d$. This would ordinarily be written a,bcd, and it represents *any* four-digit number.

We can make the form still more general, by not restricting ourselves to four-digit numbers. We can write an expression that represents any number consisting of any number of digits. To do this we need two techniques:

1. We indicate powers of any number by writing a small figure above and to the right of it. We have already indicated the square of 7 by 7^2, where the ^{2} indicates that we have *two* 7's multiplied together, $7^2 = 7 \times 7 = 49$. In the same way 7^3 means *three* 7's multiplied together, $7^3 = 7 \times 7 \times 7 = 343$. The principle holds for any power, 7^4, 7^{23}, etc.

When we apply this to the number 10 we find that the "exponent," the number above the line, tells us how many zeroes are after the 1. For instance $10^2 = 10 \times 10 = 100$, and there are two zeroes. Then 10^4, for instance, equals four 10's multiplied together, $10^4 = 10 \times 10 \times 10 \times 10 = 10{,}000$, and there are four zeroes.

2. We introduce a symbol which means "form a summation," the symbol Σ. This is a Greek "s," and we choose the letter "s" to suggest the word "sum." An example shows how it is used:

$$\sum_{n=1}^{n=3} 2^n = 2^1 + 2^2 + 2^3$$

Now we put these two ideas together and we can write the most general number. First, let us look again at the four-digit number $a \cdot 1{,}000 + b \cdot 100 + c \cdot 10 + d$. Using the power notation, we write this as $a \cdot 10^3 + b \cdot 10^2 + c \cdot 10^1 + d \cdot 1$. (In the last term d is multiplied by 1, which has no zeroes after it.) The powers of 10 are in a suitable form to use with the Σ symbol, because we can indicate all of them at once by the symbol 10^n. The exponent n takes the values 3, 2, 1, and zero in turn, in the four-digit number. But we must replace the *a, b, c, d* by new letters, like this:

$$a = a_3$$
$$b = a_2$$
$$c = a_1$$
$$d = a_0$$

Then we can write:

$$a \cdot 1{,}000 + b \cdot 100 + c \cdot 10 + d = \sum_{n=0}^{n=3} a_n \cdot 10^n$$

We can now go over to the most general number, say a k-digit number, where k may have any value we wish. We do this by writing our most general number, say N, in this form:

$$N = \sum_{n=0}^{n=k} a_n \cdot 10^n$$

By using this symbol and following the same lines of reasoning as in the preceding sections, we can derive our rules of procedure for numbers of any desired length.

Other procedures of the Trachtenberg method, besides those mentioned above, can be derived by methods of the same general nature.

CHAPTER EIGHT

Postscript

This then is the Trachtenberg System of Basic Mathematics, an entirely new approach to the very important arithmetical skills. If you have applied yourself diligently you should at this point have at least a minimum working grasp of this new mathematical system. No doubt there were moments when the going seemed difficult, but that is quite natural. The Trachtenberg system is different from what you have been accustomed to, and it is never easy to break down the habits of the past and the old ways of doing things. A little patience will overcome this, and the benefits will repay the effort many times over.

The value of this new approach will be most apparent to those who teach arithmetic to the primary grades. As we saw in the first chapter, we are no longer compelled to dull the natural eagerness of young minds with simple, prolonged repetition. The old system forces them to spend several years memorizing the multiplication tables. This means memorizing countless combinations of numbers, each one meaningless in itself, and the natural result is a feeling of boredom. The whole matter becomes distasteful. With the new system, however, we are able to keep the subject alive, and natural interest carries the young student ahead.

In this book, of course, the subject has been presented from the point of view of the adult; when it is presented to children it needs to be done somewhat differently, with changes in the emphasis on various points. This is an interesting subject, but limitations of space prevent us from going into the details here.

The important point is that all this has actually been taught to children. Since the late nineteen-forties one group after another has entered the Trachtenberg Institute and gone through the course of study. In fact, it began even earlier, when Professor Trachtenberg himself gave individual instruction to several children. Then he founded his Institute and the teaching was done in the form of classes, with assistant instructors helping him. In this way they developed the details of the method of instruction and worked them out to best advantage, over a period of some years.

From the beginning, however, the results were gratifying. The students were always fascinated by their newly-acquired powers, and their eagerness kept them going ahead. Instead of being repelled by monotony, they were attracted by the diversity of the ideas. Step by step their interest was kept alive by their own successful achievements. We have already seen for ourselves, from the beginning of Chapter One on, how much novelty there is to attract the students. At each step there is some additional feature at the same time that there is usually some similarity to what preceded it—enough to provide a desirable continuity. It is easy to understand why those children did so well.

They did well in other subjects, too. The attitude of interest or dislike tends to strike a broad target. Those who are doing badly in one kind of study will soon dislike all studies: they will "hate school." It is a natural human feeling. In the same way those who are doing well, those who are making progress every day and gaining some new skill, go to other studies with a lively interest. They approach every subject with self-confidence and the general air of asking "what can I get out of it?" This is the right way to start, and unless something strongly repels them in the other subject they will succeed.

We would like to see this sort of thing happen in all schools throughout the nation. It is true, of course, that for many reasons changes on a large scale are always made slowly. No one can reasonably expect any important changes in our national educational methods to occur in less time than decades. But if this can be done, even gradually, what a boon it would be! The children themselves would be freed from the burden of their worst drudgery, and the subject that most of them consider most difficult would become lighter. For many of them the truly fascinating possibilities would begin to open up.

From such a start, more and more of them would begin to be attracted to the physical sciences, in which mathematics is essential, and we can hope that eventually this would help to relieve the nation's most urgent need, the need for engineers and experts in the physical sciences.

It is too late for us, the adults, to benefit from such an educational change in a direct way—we cannot enjoy the same benefits as the children. But even in the matter of children and schools we do benefit indirectly. A civilized person is a member of a community. What hurts the community hurts each member, whether he is aware of it or not, and what helps the community will in the long run bring indirect benefits even to members who are not directly concerned.

More important than these indirect effects, we grown persons can get something out of the Trachtenberg system ourselves, in a direct way. Who are the students at the Trachtenberg Institute? Are they all children? No, by no means.

There are classes for children and classes for adults. Those in the adult classes are more enthusiastic and more outspoken in their praise than the children. They are more enthusiastic because they are more aware of the value of what they are receiving.

The most obvious and immediate benefit is a very practical one: the increased skill in calculation itself. Nowadays the trend of things is such that skill and accuracy in calculating is becoming a necessity for everyone. Most of us are engaged in occupations which are not primarily mathematical, but mathematics keeps creeping in. Even the portrait-painter has to make out income-tax returns like the rest of us, so he needs to keep some records of his irregular income. The man who owns and manages a small service station may be an auto mechanic by experience and by inclination, but he must be a part-time accountant. He must keep fairly complicated records of the parts purchased and the work done, the income and social-security credits of his employees, and so on. It is much the same with all of us.

In handling these necessary transactions the speed and ease of the Trachtenberg system are a great help. In the most direct manner the new improved methods will help to reduce the time spent on such matters. This is an obvious advantage.

Less obvious, but equally important, is the stress which the new system places on accuracy. As we mentioned before, a calculation is not finished until we have the *right* answer, in fact not until we can *prove* that we have the right answer. This principle is seldom observed in everyday life. Usually the results are not checked at all; if any attempt is made it is by repeating the work, which is a weak and unreliable test. Better methods exist. Scattered through the preceding chapters are repeated references to checking the work, most often by the "digit-sum" method and the "elevens-remainder" method. Both of these are good. When used together, as a double check, they are excellent. In Chapter Four (Addition and the Right Answer) we saw a special type of check, invented for a particular operation. In Chapter Five (Division—Speed and Accuracy) there was a different kind of emphasis on getting the right answer. A whole method, the "simple" method of division, was offered for those who may find it useful, and it was included because it is very easy to learn and is so devised as to minimize the danger of error. All this emphasis on being certain there is no error is part of the Trachtenberg system. The average citizen, in everyday life, makes far too many mistakes. Something is needed to overcome this tendency. We believe this point to be so important that we have emphasized it intentionally.

Besides this checking process, which locates and corrects errors, there is found to be a different kind of increase in accuracy. Throughout the system we have been concerned with the gradual development of the power of concentration. This was done in greatest detail in Chapter 1. It continued throughout the other chapters mainly in the form of the step-by-step manner in which the work was arranged. The habit of proper concentration developed in this way is a protection against errors being made at all, and there will be fewer errors for the checking process to look for.

Finally, success in acquiring these new techniques gives the learners self-confidence. For many of them the feeling of self-confidence is something new. Many persons are somewhat intimidated by the thought of any sort of calculation; they approach the work with uneasiness, half expecting the results to be wrong. This attitude almost concedes the "victory" to the problem. When the attitude is changed, and they begin to have real, solidly-based self-confidence, things begin to improve. They do the work in the right way, keeping everything under control, and that makes it less likely that any errors will occur.

The total effect of all these factors has often been a reawakened interest in mathematics and related subjects generally. This revitalization is even more important than all the practical results of the Trachtenberg system. It is our hope that the American people will receive the full benefit of these opportunities, both practical and general. We are convinced that the Trachtenberg system will be more and more widely appreciated as time goes on.